The Good Pharma

Remembering the Medicines, Discoveries, and People Who Changed Human Life

The Good Pharma

Remembering the Medicines, Discoveries, and People Who Changed Human Life

Subba Rao Chaganti

®

PharmaMed Press

An imprint of BSP Books Pvt. Ltd.

4-4-309/316, Giriraj Lane,
Sultan Bazar, Hyderabad - 500 001.
Telangana, India.

The Good Pharma: Remembering the Medicines, Discoveries, and People who Changed Human Life
by Subba Rao Chaganti

Published by

Pharma Med PRESS
An Imprint of
BSP Books Pvt. Ltd

PharmaMed Press ®

An imprint of BSP Books Pvt. Ltd.
4-4-309/316, Giriraj Lane, Sultan Bazar, Hyderabad - 500 001.
Telangana, India.
Phone: 040-23445688
E-mail: info@pharmamedpress.net
www.bspbooks.net/www.pharmamedpress.net

ISBN: 978-81-69620-56-7 (Hardback)

A Tribute

To the Human Beings behind the Healing

To the scientists who work through the night
chasing what might heal.
To the researchers who fail a thousand times
before something works.
To the clinicians who believe in trials
when the odds are uncertain.
To the patients who say "yes" to the unknown—
for the sake of others.
To the factory workers who calibrate every dose with care.
To the health workers who carry coolers across deserts and mountains.
To the families who wait, who hope,
who fight beside those they love.
To the regulators, logisticians, and data analysts
who make the invisible possible.
To those who believe that a molecule is more than chemistry—
That it is a chance, a life, a tomorrow.
To all of you:
Thank you.
This book remembers what you do.
And so should the world.

Disclaimer

This book is written for educational, historical, and informational purposes only. It is not intended to serve as medical advice, clinical guidance, diagnosis, or treatment recommendation. Readers should always consult qualified healthcare professionals regarding any medical condition, treatment, or therapeutic decision discussed or implied in this work.

The content of this book represents the author's independent research, interpretation, and narrative synthesis of publicly available historical records, peer-reviewed scientific literature, regulatory documents, biographies, archival materials, and media reports accessible at the time of writing. While every reasonable effort has been made to ensure factual accuracy, balance, and reliability, the author and publisher make no warranties or guarantees regarding the completeness or continued accuracy of the information and assume no liability for errors, omissions, or subsequent developments.

This work is not an exhaustive account of all the discoveries, innovations, institutions, companies, scientists, clinicians, regulators, patients, or contributors involved in the advancement of pharmaceutical science and medicine. The drugs, events, organizations, and individuals discussed are illustrative and representative examples, selected to highlight major milestones and themes. There are many more contributors and contributions, both known and unknown, whose impact on global health and human well-being may not be explicitly mentioned in this book.

To enhance accessibility and narrative engagement, some accounts—including patient experiences and historical vignettes—have been anonymized, paraphrased, or carefully reconstructed from multiple public sources to clarify, maintain continuity, and enhance storytelling flow, while preserving their essential factual and ethical integrity. Any resemblance to specific living individuals in such anonymized accounts is unintentional and coincidental.

The mention of specific pharmaceutical products, brand names, companies, institutions, regulatory bodies, or individuals is for historical and educational purposes only. Such mention does not imply endorsement, sponsorship, criticism, or affiliation. This book is not sponsored, funded, commissioned, or endorsed by any pharmaceutical company, healthcare organization, or commercial entity.

Opinions, interpretations, and conclusions expressed in this book are those of the author alone and do not necessarily reflect the views or official positions of any company, institution, government body, regulatory authority, or professional organization referenced.

In preparing this book, the author may have used digital and artificial intelligence–based tools selectively to assist with language refinement, structural organization, and editorial clarity. All research, analysis, interpretation, narrative framing, and final editorial decisions are the sole responsibility of the author. No artificial intelligence system was used to originate historical claims, scientific conclusions, or proprietary content. The author retains full authorship, accountability, and responsibility for the work.

This book does not reproduce proprietary, confidential, or trade-secret information. All content is presented as original narrative nonfiction, based on facts and events that are either in the public domain or lawfully referenced through independent interpretation and synthesis.

Finally, this book seeks to restore balance to public discourse by acknowledging the positive contributions of pharmaceutical science to global health, without denying that challenges, controversies, limitations, and ethical debates have existed—and continue to exist. Readers are encouraged to approach the subject with critical thinking, openness, and an appreciation for the complexity of modern medicine.

Acknowledgements

Writing The Good Pharma has been both a labor of deep respect and a journey of rediscovery—one that reminded me of the immense, often unrecognized contributions of thousands of individuals behind every vial, tablet, and treatment we now take for granted.

First and foremost, I sincerely thank Anil Shah and Nikunjesh Shah, the publishers, whose unwavering support and vision made this book possible—both in print and digital formats. Your belief in the purpose and potential of this project was crucial from the very beginning.

A special thanks to Naresh Davergave and his team at PharmaMed Press, whose design expertise and attention to detail helped shape this work into a professional, readable, and visually engaging book. Your craftsmanship speaks for itself.

To my family—your presence is my foundation.

To my wife, Mahalakshmi, thank you for your unwavering love and constant encouragement.

To our children and their spouses—Srinivasa Phanindra and Geetha, Lavanya and Satya Shankara Aditya, Sowmya and Chaitanya—your curiosity and compassion inspire me every day.

And to our beloved grandchildren—Aditi, Eesha, Surya, and Shriya—may you grow up in a world where the good that science and humanity can do is always seen, valued, and celebrated.

I also want to thank the scientists, researchers, clinicians, public health leaders, and patients whose stories bring this book to life. Their breakthroughs, sacrifices, and resilience are the heart of this story.

Finally, to every reader who approaches this book with an open mind and a willingness to see the full story of pharma—not just what went wrong, but also what went right—thank you. You are the reason this book was written.

Let this be more than just a record of what the pharmaceutical world has accomplished, but a reminder of what it can still become.

Subba Rao Chaganti

Preface

Why this Book Exists?

There is a curious paradox at the heart of modern medicine.

We live longer than any generation before us. Diseases that once terrified entire civilizations—smallpox, polio, tuberculosis, HIV—have been defeated, controlled, or transformed into manageable conditions. Surgeries that were once fatal are now routine. Cancers that once meant certain death now come with hope, options, and time. Millions of children who would not have survived past childhood now grow old, raise families, and live full lives.

And yet, the very industry that made much of this possible is often spoken of with distrust, anger, and suspicion.

This book was born from that paradox.

Over time, the public narrative around the pharmaceutical industry has narrowed. Headlines focus on pricing controversies, ethical failures, marketing excesses, and corporate missteps. Many of these criticisms are valid. Some are necessary. Accountability matters—especially in an industry that deals with life, suffering, and trust.

But something essential has been lost along the way.

We have slowly forgotten the scientists who worked for years on a single molecule that never made it to market. The researchers who failed dozens of times before succeeding once. The clinicians who believed in trials when outcomes were uncertain. The patients who volunteered for experimental therapies, not knowing whether they would benefit, only that someone, someday, might.

We have forgotten the quiet, persistent good.

This book is not an apology for the pharmaceutical industry. Nor is it a defense brief. It is not written to dismiss criticism or sanitize history. Pharma, like any powerful human enterprise, has erred, stumbled, and at times caused harm. To deny that would be dishonest.

But it is equally dishonest to ignore what pharma has done right—what it has done extraordinarily well—and the scale of good it has delivered to humanity.

This book exists to restore balance.

It tells the stories behind the molecules that changed medicine and, in doing so, changed human destiny. It looks beyond balance sheets and boardrooms to laboratories, hospital wards, makeshift clinics, and kitchen tables where life quietly resumed because a drug worked.

Here, you will meet discoveries born of accident and persistence. You will encounter moments where science outran skepticism, where urgency accelerated innovation, and where compassion found expression through chemistry. You will see how a mold on a forgotten petri dish, a dog in a laboratory, a desperate parent, or a determined young researcher altered the course of history.

This is also a book about people.

About a child waking from a diabetic coma.

About a soldier surviving infection after injury.

About a mother watching her child outlive a disease that once had no mercy.

About generations who lived longer not because of miracles, but because of molecules.

The stories chosen here are not exhaustive. They are illustrative—representative of thousands more that could be told. Behind every drug named in these pages stand countless unnamed contributors: scientists, technicians, factory workers, regulators, logisticians, nurses, physicians, caregivers, and patients themselves. This book honors them all, even when it cannot name them individually.

At its heart, The Good Pharma is a book about memory.

About remembering that progress is rarely pure, but often profound.

About remembering that science is a human endeavor, shaped by fallibility as much as brilliance.

About remembering that while criticism sharpens institutions, gratitude sustains them.

If this book succeeds, it will not convince every skeptic. That is not its aim. Its aim is simpler, and perhaps harder: to invite readers to hold two truths at once—that pharma has caused harm, and that pharma has saved more lives than any other force in modern medicine.

Both are true.

Only one is regularly remembered.

This book is an invitation to remember the other.

Subba Rao Chaganti

Contents

Prologue: Reclaiming the Narrative

We live in an era of skepticism, and many people mistrust the pharmaceutical industry. Headlines often shout about profiteering, lawsuits, and scandals. Critics complain about the cost of drugs, their impact on policy, and marketing ethics. These issues are real, and they need careful examination.

But there is another side to the story: a quieter, humbler story that doesn't make the evening news but unfolds every day in clinics, hospitals, and homes around the world. It's the story of lives saved, pain eased, childhoods restored, and futures rewritten, not by miracles but by molecules.

This book is a tribute to that story.

It is a journey through time, tracing the footsteps of those who transformed ideas into molecules, and molecules into medicine. From the accidental discovery of penicillin to the deliberate engineering of mRNA vaccines, from insulin's miraculous beginnings to the genetic modifications of gene therapy, this is the story of the good that pharma has done and continues to do.

But it's not just about drugs. It's about people.

It's about a mother in 1922 watching her diabetic child slip into a coma, only to see him wake up again after a single injection of insulin.

It's about a man diagnosed with HIV in the '80s, who believed he had only months to live, and now celebrates his 60th birthday thanks to antiretroviral therapy.

It's about a woman with breast cancer who sees her tumor shrink after receiving a treatment targeted to her DNA.

It's about a grandfather who outlived a heart attack, a child who never contracted polio, and a boy with spinal muscular atrophy who took his first steps because science gave him a chance.

In these stories, you won't see corporate boardrooms or sales forecasts Instead, you'll discover breakthrough ideas, bold risks, sleepless nights in labs, serendipitous moments, and a deep, often unwavering belief in science's power to heal.

Good Pharma does not ignore the industry's criticisms. Instead, it asks the reader to accept two truths: pharma, like any powerful institution, has made mistakes, and has saved more lives than any other force in modern medicine.

This book is not an apology. It is not a defense.

It is an acknowledgment. A recognition. A restoration of balance.

Let us walk through the timeline of discovery. Let us meet the forgotten scientists, the brave patients, the molecules that changed medicine, and the moments that changed humanity.

Let us remember the good that pharma did. And still does.

The Image Problem of Pharma

Despite its life-saving innovations and relentless contributions to global health, the pharmaceutical industry is often viewed with suspicion, controversy, and distrust. To many, "pharma" conjures images of profit-driven companies, sky-high drug prices, aggressive marketing strategies, and close ties with regulators. Over time, this perception has become a significant public relations challenge—one that obscures the enormous benefits the industry has provided.

Where Did This Image Problem Come From?

1. **Pricing Controversies and Access Issues**

   Stories of life-saving drugs being priced out of reach for ordinary patients dominate headlines. Whether it's insulin in the U.S., cancer drugs in developing countries, or rare disease treatments that cost hundreds of thousands of dollars annually, the perception is that pharma prioritizes profit over patients. "How can a drug that costs a few dollars to make be sold for thousands?" is a question that resonates across social media and political debates.

2. **Scandals and Bad Actors**

   A few egregious examples, such as the opioid crisis caused by deceptive marketing or executives suddenly raising prices of old generics have come to symbolize the entire industry. The actions of a few have tainted many, creating a lasting link between pharma and moral compromise.

3. **Opaque Business Practices**

   Drug development, pricing models, clinical trials, and regulatory interactions are complex. To the public, this lack of transparency seems intentional, fostering suspicion. When people don't understand how decisions are made or where costs originate, they tend to assume the worst.

4. **Marketing and Influence in Medicine**

   Aggressive marketing to doctors, sponsorships of medical education, and industry-funded research have fostered a perception that science can be bought and that doctors are unduly influenced. The boundaries between education, promotion, and influence are often unclear.

5. **Mistrust in Institutions and Expertise**

   The pharmaceutical industry has become an easy target in a world that is increasingly distrustful of large institutions, from governments to tech companies. This loss of trust worsened during the COVID-19 pandemic, when rapid vaccine development clashed with deep-seated fears, misinformation, and geopolitical narratives.

The Impact of the Image Problem

- Public Resistance: Vaccine hesitancy, reluctance to accept new treatments, or suspicion of drug motives often originate not in the science but in mistrust of the companies behind them.
- Policy Pushback: Regulatory reforms, price caps, and lawsuits are increasingly driven by public dissatisfaction with the industry.
- Erosion of Scientific Legacy: The remarkable achievements of pharmaceutical science—such as cures, preventions, and quality-of-life improvements—are often overshadowed by skepticism.

Why It Matters

This image problem impacts public relations and public health. When people distrust the entities working to heal them, everyone suffers. This perception crisis has real-world effects: delayed care, preventable deaths, and obstacles to innovation.

And yet, behind the headlines and harsh critiques, there is a different, quieter truth. It's about the scientists who have dedicated decades to a single molecule, the researchers who keep pushing forward despite repeated failures, the companies that distribute billions of vaccine doses in record time, and the lives that are saved, not by chance, but by design.

This book exists to shed light on that other truth. Not to deny the challenges, but to provide a complete picture.

Why this Book had to be Written

Because gratitude has a short memory.

Amid the rush of daily news, public outrage, and polarized debates about healthcare, we often forget something crucial: the fact that the longer, healthier, and more hopeful life we enjoy today is primarily thanks to pharmaceutical science.

We no longer fear smallpox. We rarely hear about polio. HIV is no longer a death sentence. Cancers that once claimed lives in months now allow patients to celebrate decades. Millions of people with diabetes live full lives thanks to insulin. Organ transplants succeed because of immunosuppressants. Premature babies thrive because of neonatal drugs developed through years of trial, error, and perseverance.

Yet these stories are fading. They are being overshadowed by a narrative that simplifies a large, varied, and complex industry into a stereotype of greed and corruption. A few bad actors, pricing mistakes, and public miscommunication have cast a long shadow over decades of honest, impactful work.

This book had to be written to restore balance to that narrative.

Not to sanitize or defend, but to remember and honor. To retell the human stories behind the drugs that changed medicine and saved lives. Behind every tablet, vial, and line on a prescription pad lies a story of discovery: sleepless nights in labs, bold ideas surviving countless failures, a researcher refusing to give up on a compound, and a child who lived because of it.

We owe it to researchers, patients, and innovators to share these stories. We also owe it to ourselves to understand the full picture before passing judgment. The pharmaceutical industry is flawed, but its contributions to human health are unmatched in scale and significance.

This book is also for the next generation of scientists, clinicians, policymakers, and even critics. It reminds us that progress is real, that science saves lives, and that behind the chemistry, there is compassion. In an era where science is questioned and trust is fragile, we must return to the human stories that made it matter in the first place.

The Good Pharma (Did) is not a eulogy. It is a celebration—a clear-eyed, honest, and heartfelt tribute to the miracles of modern medicine and the people who made them possible.

That's why this book needed to be written.

Methodology and Storytelling Approach

This book is not a textbook. It is not an industry white paper, nor a technical catalog of pharmaceutical innovations.

It is a story-driven chronicle of life-changing discoveries, the people behind them, and the lives they impacted. While the tone is narrative and engaging, the foundation is strictly factual, drawn from historical records, peer-reviewed literature, medical archives, and verified personal testimonies.

How Stories Were Chosen

The core of this book focuses on breakthrough drugs—those that changed the standard of care, saved or extended the lives of millions, and made a lasting impact on global health. The criteria for inclusion were:

- A clear before-and-after impact on disease treatment or survival
- Scientific, regulatory, or societal breakthroughs or milestones
- Global impact on increasing life expectancy and quality of life
- The compelling human story behind the drug's discovery, development, or use

Where possible, stories of real patients, doctors, scientists, and families were included to add emotional depth and connect the science to real human experiences.

Chronological Structure, Human Lens

The story spans from the early 20th century to today, outlining the development of pharmaceuticals across six major eras.

- The Foundations of Healing
- The Post-War Revolution
- The Golden Age of Discovery
- Pharma in Crisis and Courage
- The Biotech and Personalized Medicine Era
- The Future Pharma Is Building

Each chapter highlights one or more breakthrough drugs, combining the scientific journey, personal sacrifices, serendipitous moments, and the real-world impact of those medicines.

Storytelling Style and Voice

To make science accessible and emotionally engaging, the writing blends:

- Narrative nonfiction techniques (scene-setting, character arcs, turning points)
- Dialogue excerpts were available from interviews, memoirs, or archival sources
- Human-centered vignettes: A mother sees her child survive leukemia. A war medic watching penicillin save a dying soldier. A patient with HIV returning to work and life after triple therapy.

The style is conversational yet credible, avoiding jargon while honoring scientific accuracy. Each chapter begins with a moment or a story that hooks the reader. Then, the science, development path, challenges faced, and the impact are gently unfolded.

Research and Sources

- All scientific details have been verified against trustworthy sources: PubMed-indexed papers, FDA archives, WHO reports, pharma company histories, biographies of researchers, and interviews.
- Patient stories were gathered from news features, oral histories, documentaries, and memoirs, and when necessary, anonymized and carefully reconstructed to protect privacy and dignity.

Limitations and Scope

This book doesn't aim to cover every drug or innovation—only those that caused major shifts in medical history. Similarly, while stories are chosen to showcase diversity across geography, disease, and demography, this isn't a statistical analysis; it's a narrative of pharma's greatest achievements.

PART I: THE FOUNDATIONS OF HEALING (PRE-WORLD WAR ERA)

Part I

The Foundations of Healing (Pre-World War Era)

The Early Breakthroughs (1890s–1940s): When Medicine Took Its First Giant Leaps

At the turn of the 20th century, medicine stood at the threshold of transformation. Diseases that had long terrorized humanity—diabetes, bacterial infections, and inflammation—still claimed countless lives. Treatments were crude, uncertain, and often fatal.

But in just a few decades, science found its footing.

This was the age when pain met chemistry in the form of aspirin, when diabetes met its first lifeline in insulin, and when a chance mold led to the most powerful antibiotics the world had ever seen. These weren't just pharmaceutical events—they were turning points for civilization.

Part I traces the origins of modern medicine, where unlikely discoveries, daring scientists, and desperate need converged to create the foundations of pharmaceutical healing.

These are stories of accident and intention, hope and hardship, and how the earliest breakthroughs began redefining life and survival.

Chapter 1

The Birth of Modern Medicine

When Molecules met Meaning, and Medicine became Measured

For most of human history, medicine was more art than science. Healing practices were passed down through tradition, intuition, and ritual. A willow bark here, a poultice there, prayers whispered, charms worn, bloodletting performed—all with varying degrees of mystery and hope. Illness was often seen as a punishment, a curse, or a test of faith. And though some remedies worked, no one quite knew why.

But something began to shift in the late 19th and early 20th centuries. A quiet revolution was brewing in the backrooms of universities and pharmaceutical workshops. Chemistry, which had long been viewed as a pursuit of alchemists and eccentrics, was now becoming precise. Biology was gaining clarity. Microscopes revealed invisible worlds teeming with bacteria. The language of molecules began to make sense.

This was the birth of modern medicine—when healing stepped out of the shadows and into the light of evidence, experimentation, and efficacy.

A New Kind of Healer

The modern physician is no longer just a comforting presence but a problem-solver. The modern scientist no longer observes nature alone—he now shapes it. This is when medicine stopped treating symptoms in isolation and began understanding disease mechanisms.

And central to this transformation was the idea that specific chemicals could be designed or discovered to alter biological processes—diseases had molecular roots, and molecules could fix them.

The Rise of the Pharmaceutical Industry

While apothecaries and herbalists existed for centuries, the pharmaceutical industry, as we know it today, began to develop in the laboratories of Germany, France, Switzerland, and the United States.

- In Germany, companies like Bayer and Hoechst began synthesizing compounds on an industrial scale, moving from dyes to drugs.
- In Switzerland, firms like Sandoz and Ciba saw the promise of pharmacology.
- In the U.S., pharmacy schools started formalizing education, and the government slowly began introducing regulations to protect consumers.

Suddenly, there was a pipeline—from molecule to medicine, lab bench to bedside.

Aspirin: The First Household Drug

This era's first blockbuster wasn't for a deadly disease, but for something more universal: pain.

When Bayer chemist Felix Hoffmann synthesized acetylsalicylic acid in 1897, he unknowingly created the world's most widely used drug: Aspirin.

But the magic wasn't just in its ability to relieve pain and fever but in its reliability, reproducibility, and scale. For the first time, a drug could be made in factories, packed in boxes, dosed precisely, and sold around the world. It marked the beginning of standardized medicine—the belief that what worked for one could work for many, safely and predictably.

The Scientific Method Arrives in Medicine

This period also made clinical trials rise, rudimentary but increasingly rigorous. Anecdotal success was no longer enough; drugs needed proof. They were tested, compared, and refined. Control groups emerged. Dosage became data-driven. Medicine became measurable.

This was radical. For centuries, healing had been subjective. Now it was entering the realm of science—not perfectly, not without resistance—but undeniably.

From Healers to Heroes

With this shift came a new kind of heroism. Scientists were no longer seen as isolated academics but as potential saviors of humanity. They tackled invisible enemies—bacteria, viruses, enzymes—and they did it with test tubes, trial designs, and stubborn hope.

And their breakthroughs didn't just lengthen lives—they redefined what it meant to be alive.

- To survive childbirth
- To outlive infection
- To breathe through asthma
- To walk despite arthritis
- To manage chronic disease rather than succumb to it

Modern medicine didn't just save bodies. It gave people their futures back.

The Age of Therapeutic Certainty Begins

By the time the 20th century hit full stride, medicine had moved from speculative and scattered to systematic and structured. Drugs had mechanisms, doctors had protocols, and patients had hope, relief, and recovery.

This was the true birth of modern medicine.

The chapters that follow in this book are born from this legacy. They tell the stories of how pharmaceutical science, grounded in curiosity, compassion, and chemistry, transformed the human condition—not with magic but with molecules—and, with it, changed the world.

From Herbs to Chemistry

Aspirin and the Dawn of Pain Relief

Long before science identified it, everyone understood pain. It was the cry of a child with a fever, the wince of a farmer's aching joints, the silent suffering of a woman with cramps or a man with a splitting headache. It was universal—and largely untreatable.

For most of human history, nature was the pharmacy. And when it came to pain and fever, the willow tree was a trusted healer.

The Bark that Soothed

Centuries before laboratory coats and chemical patents, people in ancient Egypt, Sumeria, and China boiled the bark of the white willow to make a bitter brew for treating inflammation. Hippocrates, the father of medicine, recommended it for women in childbirth. Indigenous tribes in North America used it in rituals and remedies. Unknown to them, the active ingredient was salicin, a compound with remarkable properties—though its use was intuitive, not scientific.

But nature's gift came with a price. In its raw form, salicin (later converted to salicylic acid) was harsh. It upset the stomach. It damaged the lining of the gut. It worked, but it wounded.

By the 1800s, salicylic acid was being produced synthetically and sold across Europe. Chemists had begun to extract and refine plant compounds. The era of pharmacognosy—studying drugs from natural origins—had begun. Yet the problem remained: salicylic acid cured pain, but at a steep cost to the body.

And that's when chemistry intervened.

A Chemist. A Sick Father. A Search for Something Better.

In 1897, in the German town of Elberfeld, a young chemist named Felix Hoffmann was working at the Bayer company. His father, tormented by severe arthritis, found some relief in salicylic acid—but the drug also wreaked havoc on his stomach. Desperate to ease his father's pain, Hoffmann began experimenting in the lab, trying to modify the compound to make it gentler.

What he created was acetylsalicylic acid—a molecule that retained the benefits of salicylic acid but with far less irritation. When his father took the modified version, the pain eased—and his stomach didn't rebel.

The company named it Aspirin (the "A" for acetyl, "spir" from Spiraea ulmaria, the plant from which salicin was first derived, and "in" as a popular suffix in drug naming at the time). They didn't know it then, but they had just changed medicine—and history.

From Cure to Household Name

Bayer began marketing Aspirin in powder form in 1899. Soon, it became a staple in every doctor's kit, every soldier's bag, and eventually every home. It was affordable, reliable, and astonishingly versatile—used not only for pain and fever but for arthritis, rheumatism, and later, cardiovascular protection.

In the trenches of World War I, soldiers took Aspirin to dull the agony of war. In the 1918 flu pandemic, it was dispensed widely—even as debates raged about dosage and safety. By the 1920s, it was being sold as a tablet—ushering in the era of mass-produced, over-the-counter medicine.

Aspirin's Second Act: The Heart Saver

Decades later, researchers discovered that Aspirin didn't just relieve pain—it had anti-platelet effects, meaning it could reduce blood clots. Suddenly, it wasn't just about comfort—it was about survival. Aspirin became a life-saving drug for patients with heart disease, strokes, and those at risk of cardiac events.

For a drug that began as a bark infusion, it now had the power to prevent death.

The Legacy of Aspirin

Today, Aspirin remains one of the most widely used medications in the world. Its formula is taught in high school chemistry classes. Its mechanism is studied in pharmacology courses. Its impact is carved into the very architecture of modern medicine.

But behind that familiar white tablet lies something deeper—a human story of suffering, empathy, and ingenuity.

- A father in pain.
- A son with knowledge and love.
- A molecule with the power to soothe billions.

Aspirin marked the dawn of modern pain relief.

It was the first time science didn't just replicate nature—it refined it. It transformed healing from trial-and-error to targeted chemistry. And it showed the world that a small, tasteless pill could carry the weight of relief, the hope of healing, and the beginning of something far greater.

This was not just the birth of Aspirin. It was the birth of pharmaceutical medicine as we know it.

The Real Story: The German Chemist and the Forgotten Father of Aspirin

Credit, Controversy, and the Quiet Genius behind the World's Most Famous Drug

Every great drug has two stories: the scientific one that's recorded in journals—and the human one that lives in the shadows. The story of Aspirin is no different. While the world remembers Felix Hoffmann as the man who gave us acetylsalicylic acid, fewer know the name Arthur Eichengrün—the man many believe was the real father of Aspirin.

And his story is one of inspiration, innovation—and injustice.

Felix Hoffmann: The Official Credit

In the official history books and Bayer's archives, Felix Hoffmann is credited with synthesizing a more tolerable form of salicylic acid in 1897. He did it, the story goes, to help his ailing father who was suffering from debilitating arthritis. The modified compound—acetylsalicylic acid—was effective and far less harsh on the stomach. Bayer patented it in 1899 and branded it Aspirin.

The drug went on to become a global sensation. Hoffmann received accolades and a prominent place in pharmaceutical history.

But that story, as it turns out, may be incomplete.

Arthur Eichengrün: The Forgotten Architect

Arthur Eichengrün was a senior Jewish chemist at Bayer at the same time Hoffmann worked there. A brilliant and methodical scientist, Eichengrün oversaw many of the company's early pharmaceutical projects. According to documents and his own testimony decades later, it was he who directed Hoffmann to synthesize acetylsalicylic acid as part of a broader research initiative into improved therapeutics.

In a paper written in 1949—after World War II had ended—Eichengrün claimed that he had conceptualized the drug, planned the experiments, and even selected Hoffmann to carry them out under his supervision. Hoffmann, he said, was a competent chemist but not the intellectual author of the idea.

His claims were ignored. Why? Because by then, Eichengrün had been erased from official German science narratives.

The Shadows of Anti-Semitism

During the Nazi era, Eichengrün, being Jewish, was stripped of his status, his property, and eventually his freedom. He was imprisoned in the Theresienstadt concentration camp. His contributions, like those of many Jewish scientists, were either minimized or intentionally scrubbed from history.

After the war, he tried to reclaim his scientific legacy, submitting his case to Bayer and German authorities. But he died in 1949—largely unacknowledged.

It wasn't until 1999, a full century after Aspirin was created, that Walter Sneader, a researcher at the University of Strathclyde in Scotland, re-examined the historical evidence. After analyzing lab records, internal memos, and timelines, Sneader concluded that Eichengrün, not Hoffmann, deserved credit as the intellectual father of Aspirin.

Bayer rejected the claim. But the debate was reopened.

What the Records Suggest

- Hoffmann synthesized acetylsalicylic acid—but may not have fully understood its pharmacological potential.
- Eichengrün proposed the idea, supervised the synthesis, and advocated its clinical testing and commercialization.
- Eichengrün later developed several other important drugs and pioneered early work on cellulose acetate, a plastic that revolutionized industrial materials.

Yet his name never appeared on the Aspirin patent.

Legacy and Recognition

Today, most historians agree that Aspirin's discovery was collaborative—but that Eichengrün's role was unjustly erased due to the politics and prejudices of the time.

Felix Hoffmann will always be remembered for the synthesis. However, Arthur Eichengrün, the quiet genius, deserves recognition as the person who saw the potential, directed the process, and perhaps, gave the world its first true modern drug.

His story is a reminder that science is not always fair, that behind every drug lies a tangle of personalities, politics, and persistence—and that sometimes, history needs correcting.

As we celebrate what Aspirin did for the world, we must also remember who made it possible, even if the world tried to forget him.

Chapter 2

Insulin: A Miracle for the Sugar Disease (1921)

The Dog, the Boy, and the Discovery that Resurrected Lives

Before 1921, a diagnosis of diabetes mellitus—especially Type 1 diabetes—was a slow-motion death sentence.

Known commonly as the "sugar disease," it wasted the body away from the inside. Without insulin, patients couldn't metabolize glucose, their primary source of energy. The sugar stayed trapped in the bloodstream, poisoning organs while the body starved at the cellular level.

Children were the most tragic victims.

Families watched helplessly as their once-healthy sons and daughters became skeletal, listless, and lost. Doctors, desperate to buy more time, prescribed starvation diets: a few hundred calories a day—barely enough to survive, and certainly not enough to grow. It was cruel compassion. It prolonged life but didn't prevent death.

Then came Insulin. And with it, one of the most dramatic turnarounds in the history of medicine.

The Breakthrough No One Expected

In the summer of 1921, in a small laboratory at the University of Toronto, a 29-year-old surgeon with no formal research background walked in with a wild idea.

His name was Dr. Frederick Banting.

He believed that the pancreas produced a substance that regulated sugar in the blood, and that if he could extract it—pure and uncorrupted—he could treat diabetes. Many researchers had tried and failed. Banting wasn't even an endocrinologist. But he had determination, a crude surgical plan, and one key supporter: Professor J.J.R. Macleod, who gave him lab space, 10 dogs, and a bright medical student assistant: Charles Best.

The Dog that Changed the World

They began their experiments in July 1921. The plan was to ligate (tie off) the pancreatic ducts in dogs, allowing the enzyme-producing cells to die off while preserving the insulin-producing islets of Langerhans. Then, they would extract a pancreatic "juice" from the treated dogs and inject it into diabetic dogs to see if blood sugar would drop.

Most dogs died.

But one—a diabetic dog named Marjorie—received the extract and lived. Her blood sugar dropped dramatically. The team had proof: they had isolated something powerful—something life-saving.

By early 1922, they had enough confidence to try it in humans.

The Boy in the Hospital Bed

The first patient was Leonard Thompson, a 14-year-old boy admitted to Toronto General Hospital. His diabetes was advanced. He was pale, gaunt, and drifting in and out of consciousness. Weighing barely 65 pounds, he was days from death.

On January 11, 1922, he received the first injection of raw pancreatic extract.

It only partially worked—and caused a serious allergic reaction due to impurities. But the team, now joined by biochemist James Collip, rushed to purify the extract. Twelve days later, they tried again.

This time, it worked.

Leonard's blood sugar stabilized. He regained energy. The change was almost immediate—as if life had returned to his body.

The Miracle Spreads

Word spread like wildfire. Soon, other children with diabetes were being injected. Entire wards once filled with dying children transformed into halls of recovery. Parents wept. Nurses cheered. Doctors wrote in disbelief.

It was more than a treatment—it was a resurrection.

By the end of 1923, insulin was being mass-produced by Eli Lilly, thanks to an agreement with the University of Toronto. That same year, Banting and Macleod were awarded the Nobel Prize in Physiology or Medicine—Banting, furious that Charles Best wasn't recognized, later shared his prize money with him.

Insulin was the first clear case in medical history where a previously fatal disease became manageable overnight.

Why it Mattered So Much

Insulin was more than just a new drug. It was a proof of concept:

- That hormones could be isolated and administered
- That chronic diseases could be managed with targeted therapy
- That a small research team with limited resources could change the world

And it ushered in a new era of pharmaceutical science—one focused not just on treating symptoms but on replacing what the body lacked.

The Human Story Lives On

Since that day in 1922, insulin has saved hundreds of millions of lives. It's given children a future, adults a chance at parenthood, artists their creativity, scientists their careers, and ordinary people their normalcy.

Yet despite its life-giving power, insulin's story isn't over. In recent years, its affordability has become a global issue—a reminder that even miracles need good stewardship.

But this chapter, this moment in history, remains unshakable:

One boy. One dog. One determined doctor.

And a discovery that turned despair into hope, and death into life.

The Story of Banting, Best, and the Dog That Saved Millions

How Grit, Guts, and One Loyal Dog Led to One of Medicine's Greatest Miracles

It began not in a grand research institute, but in a cramped laboratory, in the sweltering summer of 1921. The lab was hot, under-equipped, and filled with barking. It smelled of sweat, disinfectant, and determination.

And in that lab, a surgeon who couldn't let go of an idea, a medical student with relentless energy, and a series of diabetic dogs would unwittingly rewrite the history of medicine.

One of those dogs—just one—would become the key to saving millions of lives. This is that story.

The Surgeon with an Idea

Dr. Frederick Banting wasn't famous. In fact, he was struggling. A young Canadian physician, he was giving lectures to keep himself financially afloat when, one night in October 1920, he came across a journal article that sparked an idea.

It said that the pancreas had a group of cells—the islets of Langerhans—which seemed to influence sugar metabolism. Banting's mind lit up. What if the digestive enzymes of the pancreas were destroying this mysterious internal secretion before it could be extracted? What if he could remove or disable the enzyme-producing parts and isolate the substance the body lacked in diabetes?

He jotted a note in the middle of the night:

"Ligate pancreatic ducts of the dog. Keep dogs alive till acini degenerate, leaving islets. Try to isolate the internal secretion of these and relieve glycosuria."

It wasn't elegant. But it was a blueprint for the impossible.

The Reluctant Professor and the Enthusiastic Student

When Banting pitched the idea to Professor J.J.R. Macleod at the University of Toronto, the professor was skeptical. The plan was crude, the risks high, and Banting had little research experience. But perhaps to be rid of him—or perhaps out of curiosity—Macleod gave Banting a lab, 10 dogs, and an assistant.

That assistant was Charles Best, a 22-year-old medical student with keen instincts and boundless energy. From that moment, Banting and Best became inseparable—working 16-hour days, doing their own surgeries, drawing blood, running tests, and keeping detailed logs.

Their goal? Induce diabetes in dogs by removing their pancreas, then try injecting them with extracts from healthy pancreatic islets.

The Dogs that Made History

The early experiments were brutal. Most of the diabetic dogs died from infections, surgical complications, or malnutrition. But the pair kept going—driven by purpose, not prestige.

Then came a dog known only as Dog #408—nicknamed Marjorie.

She had been made diabetic and, without insulin, would have died within days. But Banting and Best injected her with their latest pancreatic extract. Her blood sugar dropped. She became alert again. The signs were unmistakable.

For 70 days, Marjorie lived on daily injections, her diabetes controlled by the crude extract that the world would one day know as insulin.

She was the first diabetic mammal to survive on insulin.

The Human Leap: Leonard Thompson

With Dog #408's success behind them, the team was ready to try the extract on a human. The patient was Leonard Thompson, a 14-year-old boy dying of diabetes at Toronto General Hospital.

The first injection was disappointing—impure, and it caused an allergic reaction. But a new team member, biochemist James Collip, quickly refined the extract. On the second try, Leonard's blood sugar plummeted, and his energy returned.

He became the first human successfully treated with insulin.

The Legacy of Dog #408

Without Marjorie—Dog #408—there would have been no human trial. No Leonard Thompson. No millions of lives saved. Her body processed those first imperfect extracts. She endured and responded, giving Banting and Best the proof they needed to believe.

They didn't know it at the time, but they had changed the future—not just of diabetes, but of medicine itself. Insulin became the first hormone ever successfully extracted and used therapeutically. It marked the beginning of replacing missing substances in the body with purified biologics—an idea that would fuel generations of life-saving innovation.

Recognition, Controversy, and Immortality

In 1923, Banting and Macleod received the Nobel Prize for their discovery. In a gesture of fairness, Banting gave half his prize money to Best. Macleod, in turn, shared his with Collip. Politics and personality aside, what they created together was undeniable: a new lease on life for people who once had none.

And though Marjorie didn't live long enough to see what her sacrifice made possible, her place in history is sacred.

She wasn't just a lab animal.

She was the bridge between theory and miracle.

Today, insulin keeps over 150 million people alive.

And it all began with Banting, Best, and the dog that saved millions.

How Insulin Changed Lives Forever

From Death Sentence to Decades of Life: The Drug that Gave the World a Second Chance

Before 1921, a diagnosis of Type 1 diabetes was a slow farewell. The disease didn't kill suddenly—it stole a little more each day: weight, energy, appetite, strength, and finally, life. The only known "treatment" was a starvation diet, sometimes as low as 300–400 calories a day, which kept blood sugar levels from spiking—but at a terrible cost. Children would fade into skeletal outlines of themselves. Parents were told to make peace. There was no real hope.

And then insulin arrived.

It didn't just add years to lives—it redefined what was possible for people with diabetes. For the first time in history, a diagnosis of diabetes didn't mean preparation for death. It meant planning for life.

From Coma to Conversation: The Leonard Thompson Effect

In early 1922, 14-year-old Leonard Thompson was unconscious, slipping into a diabetic coma. After receiving purified insulin, he woke up. His energy returned. His appetite came back. His body, moments from collapse, rallied.

That transformation—stark and immediate—sent a signal around the world: this changes everything.

Wards of the Dying Became Halls of the Living

Hospitals quickly filled with diabetic children brought in by desperate parents. In scene after scene, doctors injected the new extract. What followed was nothing short of resurrections. Pale faces flushed with color. Hollow eyes lit up. Children began to sit up, ask for food, and speak again.

In one ward in Toronto, dozens of children who were slowly wasting away were treated with insulin. Within days, the ward that resembled a hospice began to look like a playground.

The transformation wasn't metaphoric—it was visibly miraculous.

The First Generation to Live With Diabetes

Thanks to insulin, children who once didn't survive past puberty grew up. They went to school, fell in love, had families, built careers. Insulin gave them not just survival—but normalcy.

People with diabetes became pioneers of a new reality:

- A girl who would have been buried by age ten now becomes a schoolteacher.
- A boy whose diagnosis came with mourning becomes a grandfather.
- A young artist with insulin at hand paints a whole life's work.

Their stories became living proof of what science could do.

Athletes, Mothers, Inventors: Breaking the Boundaries of Disease

With access to insulin and proper care:

- Olympic medalists were born.
- People with diabetes climbed mountains, ran marathons, flew planes.
- Women with diabetes carried pregnancies to term—previously impossible.
- Engineers, scientists, artists, authors—all proved that diabetes no longer defined destiny.

Insulin gave people control over their lives, not just their blood sugar.

The Global Ripple Effect

The impact of insulin went far beyond the West. As its production scaled globally, it saved lives in countries across Africa, Asia, Latin America, and Eastern Europe, where access to modern medicine was limited.

Insulin became part of the UN list of essential medicines. It helped build trust in pharmaceutical science, taught doctors to manage chronic disease long-term, and opened the door to other hormone therapies—from thyroid replacement to reproductive health.

The Rise of Self-Management and Empowered Patients

Insulin was not just a drug—it created a new kind of patient: informed, responsible, and empowered.

It led to:

- The first home glucose monitors
- Patient education programs
- The birth of diabetes advocacy and support groups

Patients learned to count carbs, understand their own bodies, and actively manage their disease. A new philosophy emerged: chronic illness could be lived with rather than suffered from.

More than Medicine: A Symbol of what Science can do

Insulin remains one of the most powerful symbols of pharmaceutical hope:

- It transformed a universally fatal disease into a manageable condition
- It inspired future hormone therapies
- It laid the groundwork for biotech and biosynthetic drug manufacturing
- And it proved that molecules, when matched with compassion and grit, can perform miracles

And Yet, a Cautionary Note

While insulin has saved hundreds of millions of lives, its access remains uneven. In some parts of the world, families still ration insulin. In others, high prices cause suffering that science already solved.

This duality—of miraculous discovery and market barriers—reminds us that the responsibility doesn't end with invention. It continues with access, affordability, and advocacy.

The Legacy

Insulin didn't just change how we treat diabetes.

It changed how we think about disease.

It changed what people dared to hope for.

It changed the role of pharmaceuticals in shaping human destiny.

It took a wasting child and gave him not just days, but decades.

And in doing so, it gave the world the first great victory of modern medicine.

Chapter 3

Sulfa Drugs: The First Antibiotics (1930s)

The Red Dye that Defeated Death and Opened the Door to Modern Infection Medicine

Before the 1930s, even the smallest wound could become a death sentence.

A scraped knee, a sore throat, a surgical incision—any could lead to rampant bacterial infection. Strep, staph, and pneumonia were killers lurking behind every minor illness. Millions died not from complex diseases, but from simple infections the body couldn't fight.

Mothers wept as children succumbed to fevers. Soldiers died more often from infections than from bullets. Women died after childbirth due to puerperal fever. Hospitals were places of both healing and helplessness.

That changed with the discovery of sulfa drugs—the world's first commercially successful antibiotics.

And, as with many breakthroughs in pharma history, it started with a mistake, a hunch, and a little bit of red dye.

The Accidental Breakthrough at Bayer

In the early 1930s, German chemist Gerhard Domagk was working in the research labs of Bayer, then a division of IG Farben. The company was experimenting with synthetic dyes—not for clothing or paper, but for their potential to bind to bacteria. Domagk believed that some dyes might selectively kill pathogens without harming human cells.

One compound, a bright red dye called Prontosil, showed unusual promise. In test tubes, it didn't seem to do much. But when Domagk tried it in live mice infected with deadly streptococcus, the results were astonishing.

The infections vanished. The mice lived.

So did infected rabbits. So did treated animals across repeated studies. Something in Prontosil was neutralizing the bacteria inside the body.

The Girl with the Needle-Pricked Finger

Then came the real-world test.

In 1935, Domagk's own six-year-old daughter accidentally pricked her finger on a sewing needle. It became infected—quickly and severely. Her arm swelled. Gangrene set in. Amputation was recommended.

Desperate, Domagk gave her Prontosil.

She survived. No amputation. No lingering infection.

This wasn't just a father saving his daughter. It was a proof-of-concept that would go on to save millions.

Unlocking the Secret: It wasn't the Dye itself

Soon, French researchers working independently found that the active ingredient wasn't Prontosil's color or dye properties, but what it broke down into inside the body: a molecule called sulfanilamide.

This compound, cheap and easily synthesized, became the foundation of the sulfa drug revolution.

The sulfonamides were the first true antimicrobial agents—able to kill or stop the growth of bacteria inside the human body without harming the host.

The Revolution in Hospitals

By the late 1930s, sulfa drugs were being widely prescribed around the world:

- Strep throat, scarlet fever, meningitis, and wound infections became treatable.
- For the first time, doctors had power over bacterial disease, not just guesses.
- Post-surgical infections dropped. Childbirth-related deaths plummeted. Wartime injuries became survivable.

Sulfa drugs redefined what it meant to be a doctor. Physicians were no longer passive observers of infection—they were combatants. And patients? They began surviving what once killed indiscriminately.

The Human Stories behind the Molecule

One of the most famous cases was that of Winston Churchill, who, in 1943, came down with a severe pneumonia while visiting North Africa during WWII. He was treated with a sulfa drug and recovered fully. He later said, "Sulpha drugs gave me back my life."

Another pivotal moment: During the 1936 meningitis outbreak in the U.S., hundreds of children were saved thanks to the swift deployment of sulfanilamide. Hospitals that once delivered death notices now deliver recovery.

The Tragedy that Prompted Regulation

But the sulfa story also gave rise to one of the most tragic moments in American pharmaceutical history—and led to the creation of modern drug regulation.

In 1937, a U.S. manufacturer prepared an elixir of sulfanilamide using diethylene glycol—a toxic solvent. Over 100 people, mostly children, died of kidney failure after taking it. The disaster led directly to the 1938 Food, Drug, and Cosmetic Act, which required drugs to be tested for safety before being sold.

This act laid the groundwork for the FDA's regulatory authority—ensuring that drug safety could never again be taken for granted.

The Legacy of Sulfa Drugs

Sulfa drugs were the first in a long line of antibiotics. They saved millions of lives before penicillin became widely available. And even today, sulfa-based compounds remain essential in treating urinary tract infections, pneumonia in immunocompromised patients, and certain forms of meningitis.

But perhaps their greatest legacy is symbolic:

They proved that bacteria could be beaten with chemistry, and that synthetic molecules could do what nature hadn't yet evolved. They turned medicine into a weapon—not just against symptoms, but against causes.

From Red Dye to Renaissance

It all began with a dye. With a hunch. With a girl who pricked her finger.

And from that—came a turning point in human history. The sulfa drugs opened the door that penicillin would soon burst through. But they were first. And they were proof that pharmaceutical science could bend the arc of life toward survival.

They didn't just treat infection.

They transformed expectations.

How a Red Dye Led to The First Antibacterial Wonder

The Curious Journey of Prontosil — From Textile Labs to Hospital Wards

In the early 1930s, hospitals across the world were filled with dread—not because of cancer or heart disease, but because of infections that no one could stop. A child with strep throat, a woman giving birth, a man recovering from surgery—any of them could die within days if bacteria entered their bloodstream. There were no antibiotics. No cures. Only prayers.

Yet the solution—the world's first antibacterial wonder—came not from a traditional medical lab or a university.

It came from a red dye. One that stained fabrics, not bandages.

And it changed everything.

From Textiles to Test Tubes: A Chemist's Hunch

At the German pharmaceutical company Bayer, part of the massive chemical conglomerate IG Farben, scientists were studying synthetic dyes—not for medicine, but for industry. The vibrant red, purple, and blue dyes made for eye-catching clothing, wallpapers, and inks. But a few researchers believed these dyes might have biological properties as well.

One of them was Gerhard Domagk, a pathologist with battlefield experience from World War I. He had seen firsthand how bacterial infections killed more soldiers than bullets. He carried those memories into the lab.

Domagk theorized that if certain dyes could selectively bind to fabrics, maybe they could also target bacteria without harming human cells.

It was a long shot.

The Discovery of Prontosil

Among dozens of dye compounds synthesized by Bayer chemists, one stood out: a bright brick-red azo dye called Prontosil. It had no known medical use and didn't kill bacteria in test tubes. By conventional standards, it was a dead end.

But Domagk wasn't deterred. He believed in testing compounds inside living organisms, not just in lab dishes. He infected mice with deadly streptococcus and injected them with Prontosil.

To his astonishment, the mice survived.

Again and again, in study after study, Prontosil protected animals from bacterial infection. Domagk had discovered the first systemically effective antibacterial drug in history.

And it had come from a dye.

A Child's Life Saved: The Daughter Who Didn't Lose Her Arm

The true test came in Domagk's own home. In 1935, his six-year-old daughter, Hildegarde, pricked her finger while sewing. The wound became infected, and gangrene began to spread rapidly. Doctors advised amputation.

Domagk refused. He gave her an injection of Prontosil.

She recovered completely, her arm intact.

It wasn't just a medical victory—it was deeply personal. The red dye had saved his daughter's limb—and likely, her life.

Unmasking the Active Ingredient

Across the border, in France, scientists at the Pasteur Institute wanted to know why Prontosil worked. They discovered that once inside the body, the dye broke down into a colorless compound called sulfanilamide.

It was this molecule—not the dye itself—that killed bacteria.

This revelation opened the floodgates. Sulfanilamide could be easily synthesized without the dye. It became the first mass-produced antibacterial drug—the cornerstone of the sulfa drug era.

A New Era in Medicine

Doctors suddenly had a weapon against diseases like:

- Pneumonia
- Childbirth infections
- Meningitis
- Wound sepsis
- Strep throat and scarlet fever

Infections that once killed millions were now treatable.

In 1936, when the son of U.S. President Franklin D. Roosevelt developed a severe strep infection, he was saved by sulfa drugs. The medicine gained not only trust—but global fame.

Why it Mattered

The discovery of Prontosil did more than save lives. It proved that:

- Synthetic chemistry could be used to fight disease
- Drugs didn't need to come from nature—they could be designed
- The chemical structure of a molecule could be manipulated to target pathogens
- Pharmaceuticals had moved beyond pain relief and symptom control—into the realm of curing infections

It was the first true antibacterial revolution—years before penicillin reached mass production.

From Red Dye to Redemption

Prontosil didn't just stain fabric—it stained the history of medicine with hope. It was the red thread that tied together chemistry, courage, and compassion.

It showed the world that molecules could be engineered to heal, and that sometimes, the answer to a medical mystery doesn't come from where we expect—but from where we dare to look.

A Child's Life Saved, A Nation Grateful

How Sulfa Drugs Turned Grief into Gratitude and Earned the World's Trust

In the winter of 1936, in the town of Circleville, Ohio, a small boy lay feverish and fading. His body burned with infection, his throat swollen shut. He had streptococcal sepsis, a diagnosis that, at the time, meant only one thing: prepare for the worst.

His parents sat by his bedside, helpless, like so many before them, watching their son drift further from them hour by hour. But what happened next would not only save their child—it would ripple across a nation, solidify trust in pharmaceutical science, and change how we confront disease forever.

That boy was Franklin Roosevelt Jr., the son of the sitting President of the United States.

The White House Emergency

When the President's son fell gravely ill, the country held its breath. In an era where the death of a child from a throat infection was tragically common, the nation feared losing not only a child but a symbol of its First Family.

Medical experts rushed to his bedside. Traditional methods failed. The infection spread rapidly. Time was running out.

Then, a young doctor suggested an experimental treatment: sulfanilamide, a derivative of the newly discovered Prontosil—the first sulfa drug, the first systemic antibiotic.

The stakes were high. The drug was new. No one was sure if it would work in this case.

But with little to lose, the injection was administered.

A Dramatic Turnaround

Within hours, the boy's fever began to drop. His breathing eased. His bloodwork improved. And within a day, he sat up in bed, weak but alert—alive.

The White House released the news: The President's son had been saved by a sulfa drug.

It was the first time a sitting president had publicly embraced an experimental therapy, and the story electrified the press. Headlines read:

"WONDER DRUG SAVES ROOSEVELT SON"
"MIRACLE MEDICINE BATTLES DEATH"

Across America, parents read the story and saw their own sons and daughters in that hospital bed. They weren't just relieved for the First Family—they were hopeful for their own.

Sulfa Drugs become a National Treasure

Almost overnight, sulfa drugs became a symbol of scientific hope. Doctors demanded supplies. Hospitals changed protocols. Governments prioritized production and distribution.

Pharmaceutical companies scaled up manufacturing. And for the first time, American medicine had a cure for deadly bacterial infections—not just treatments to soothe, but drugs to heal.

Public perception of pharmaceuticals shifted dramatically:

- What had once been seen as guesswork became precision.
- What had once been seen as passive now felt powerful.
- Science, for the first time, had shown that it could snatch life back from death's grip.

Beyond the Roosevelt Family

While the Roosevelt case captured headlines, thousands of similar stories unfolded quietly across the country:

- A young girl in Kansas with pneumonia lived because her doctor had read about sulfa drugs.
- A factory worker in Chicago survived a staph infection after a workplace accident.
- Soldiers with infected wounds in training camps were successfully treated before shipping off to Europe.

Each recovery became a silent testimony to what the Roosevelt family had experienced: sulfa drugs work.

A Nation Grateful

The rescue of Franklin Roosevelt Jr. didn't just change a life—it galvanized public trust in science and in the pharmaceutical industry. For the first time in history:

- Americans saw a drug act like a miracle.
- Doctors had a tool that truly cured, not just comforted.
- The pharmaceutical industry was seen not as a commercial force—but as a lifesaving engine of innovation.

It created a cultural shift, making people more receptive to future breakthroughs like penicillin, insulin, and vaccines.

In a moment of crisis, a child's recovery became the foundation for a nation's faith in modern medicine.

The Red Thread of Trust

From the red dye of Prontosil to the pink cheeks of a recovering boy, sulfa drugs wove a red thread of hope through history.

A life was saved.

A family rejoiced.

A nation believed.

And that belief changed the course of medicine forever.

PART II: POST-WAR PHARMACEUTICAL REVOLUTION (1940s – 1960s)

Part II

Post-War Pharmaceutical Revolution (1940s - 1960s)

From Battlefield Discoveries to Living Room Medicines: The Era That Changed Everything

When the dust of World War II settled, the world stood at a crossroads. Nations were rebuilding, science was accelerating, and hope—after years of destruction—was being stitched back into daily life.

In medicine, the war had revealed two powerful truths:

1. Infectious diseases could be defeated with the right molecules.
2. Science could move faster than ever imagined when lives were on the line.

And with that momentum, the pharmaceutical industry catapulted into its first golden age—a time when lab coats replaced luck, and drug discovery moved from chance to system.

Between the 1940s and 1960s, an explosion of pharmaceutical innovation transformed how humanity lived, healed, and aged. This was the era when the pharmacy became part of every household, and medicines extended life not just by days—but by decades.

The Penicillin Boom

Though discovered in 1928, penicillin became a full-blown global force during World War II. The Allied powers scaled up production to treat infected wounds and battlefield injuries. Soldiers who once died from gangrene or pneumonia now walked out of field hospitals alive.

After the war, penicillin was released to the public, and infection was no longer the world's greatest killer. Pneumonia, syphilis, gonorrhea, and strep infections suddenly had an enemy. Life expectancy jumped. Families who once lost multiple children to fevers now raised healthy generations.

The era of antibiotics had truly begun—and with it, a new public confidence in pharmaceutical science.

Pharma goes Industrial

Fueled by wartime manufacturing advances, pharmaceutical companies expanded dramatically. What had been small research groups turned into sprawling R&D departments. Drug discovery became a structured process involving:

- High-throughput screening
- Animal models
- Early clinical trials

Companies like Pfizer, Merck, Glaxo, Roche, and Eli Lilly emerged as global powerhouses, investing not just in production, but in innovation.

The Rise of the "Wonder Drugs"

The post-war years saw the discovery and commercialization of a wave of drugs that redefined medical care:

- Cortisone (1948): Transformed treatment for arthritis, asthma, and autoimmune conditions.
- Chlorpromazine (1951): The first antipsychotic, heralding a revolution in psychiatric care.
- Isoniazid (1952): A powerful treatment for tuberculosis, then one of the world's deadliest diseases.
- Methotrexate (1956): An early chemotherapy that showed cancer could be chemically controlled.
- Oral contraceptives (approved in 1960): Gave women unprecedented control over reproduction and ignited a cultural shift.

These weren't just new medicines—they were new freedoms.

Vaccines become Global Tools

Post-war pharma also embraced preventive medicine.

The polio vaccine—developed by Jonas Salk in the early 1950s—was tested in one of the largest medical trials in history. When the vaccine was declared safe and effective, church bells rang, and parents wept.

By the end of the 1960s, smallpox eradication was in sight. The world had proof that disease could be eliminated—not just managed.

The Birth of Consumer Confidence

This period also saw the rise of branded, packaged medicines available at local pharmacies—aspirin, cough syrups, antacids, and cold remedies. The public now saw medicine as part of daily life, not just emergencies.

For the first time:

- Patients expected relief
- Chronic illnesses became treatable conditions
- Medical progress was measured in years added to life

Pharmaceuticals were no longer mysterious. They were modern.

Safety, Regulation, and the Thalidomide Wake-up Call

But the era wasn't without its pain.

In the late 1950s, the thalidomide tragedy struck. Marketed in Europe as a sedative and anti-nausea drug for pregnant women, it caused severe birth defects in over 10,000 children.

Though never approved in the U.S. thanks to the vigilance of FDA reviewer Dr. Frances Kelsey, the global fallout led to sweeping reforms in drug testing, safety, and regulatory oversight.

From that tragedy emerged a silver lining: the modern framework of drug regulation. The world learned that powerful drugs require powerful responsibility.

The World Transformed

By the mid-1960s, the pharmaceutical landscape had fundamentally changed:

- Life expectancy soared worldwide
- Infectious diseases fell from top causes of death
- Mental health, reproductive rights, and cancer care entered the mainstream conversation
- Pharma became a pillar of post-war prosperity and public health

Legacy of the Revolution

The post-war pharmaceutical revolution proved that science could shape society, not just support it. Medicines gave people more than survival—they gave them choice, dignity, and time.

The post-war pharmaceutical revolution proved that science could shape society, not just support it. Medicines gave people more than survival—they gave them choice, dignity, and time.

From battlefield tents to living rooms, from hospitals to schoolyards, the impact of post-war pharma is still felt today.

This was not just the rise of an industry.

Chapter 4

Penicillin: The Drug that Won the War

How a Forgotten Petri Dish and a Global Crisis gave the World its First True Superdrug

It began with a mistake.

In the summer of 1928, Scottish bacteriologist Alexander Fleming returned to his cluttered lab at St. Mary's Hospital in London after a two-week holiday. He had left petri dishes filled with staphylococcal bacteria—those responsible for boils, abscesses, and even deadly sepsis—sitting on his workbench.

One dish, forgotten near an open window, had grown mold. But not just any mold. Around it, the bacteria had disappeared—killed in a perfect halo.

Most researchers would have tossed the dish. Fleming looked closer.

What he had found—purely by accident—was a mold of the genus Penicillium, and it was producing something that killed bacteria.

He called it penicillin.

The Struggle to Turn Mold into Medicine

Fleming had discovered something extraordinary, but he wasn't equipped to harness it. The mold was fragile, the compound unstable. He couldn't isolate or purify it effectively.

For over a decade, penicillin remained a scientific curiosity, noted in academic circles but without clinical application.

Then, on the eve of the Second World War, two scientists at Oxford—Howard Florey and Ernst Chain—revisited Fleming's work. Where Fleming saw possibility, they saw urgency.

Europe was heading into chaos. Wounds, infections, and death were coming. If penicillin could be tamed, it could save millions.

They assembled a team, including the often-overlooked Norman Heatley, and began a tireless process of refining and purifying penicillin. It took months of growing mold in milk bottles and bedpans, scraping, extracting, and testing.

Then, in 1941, they had a breakthrough.

The First Human Trial — And a Heartbreaking Race

Their first real patient was a London police officer with a life-threatening infection from a simple scratch. He was dying.

They injected him with purified penicillin. He began to recover almost immediately. His fever dropped. His wounds began to heal. Hope surged.

But they didn't have enough penicillin to continue treatment. Desperately, they tried to recover the drug from his urine to reuse it.

It wasn't enough.

The man died.

But now they knew: penicillin worked. They just had to find a way to produce it—fast and in volume.

America Joins the Mission

In 1941, Florey and Heatley flew to the United States. The war was intensifying, and the Allies needed medical miracles.

At a U.S. Department of Agriculture lab in Peoria, Illinois, they teamed up with American scientists and began experimenting with deep fermentation—growing penicillin in vats instead of petri dishes.

Then came an unlikely hero: a moldy cantaloupe from a local market. It carried a Penicillium chrysogenum strain that was more productive than anything seen before. The team bred and mutated it to supercharge its output.

Within months, U.S. pharmaceutical companies like Pfizer, Merck, and Squibb were manufacturing penicillin on an industrial scale.

The War-Changing Moment

By D-Day, June 1944, penicillin was in every Allied medic's kit.

Wounded soldiers no longer died from infected shrapnel wounds. Amputations dropped dramatically. Post-surgical infections that once raged in field hospitals were now rare.

Penicillin didn't just save lives—it saved legions, boosting morale and survival rates across the front lines.

It became known as "the wonder drug" and was widely credited with helping the Allies win the war—not by fighting enemies, but by defeating disease.

A Public Victory, A Quiet Controversy

In 1945, Fleming, Florey, and Chain were awarded the Nobel Prize in Medicine. But behind the curtain, the work of Norman Heatley—who engineered the extraction process—was overlooked, as was the contribution of thousands of unsung lab workers, factory staff, and even that humble cantaloupe.

Still, the global gratitude was real. Penicillin became the face of pharmaceutical possibility.

A New Era Begins

Post-war, penicillin became available to the public. It was used to treat:

- Pneumonia
- Syphilis
- Strep throat
- Gonorrhea
- Skin infections
- Rheumatic fever

Hospitals saw death rates plummet. The fear of infection—which had hovered over every surgery, every childbirth, every minor injury—dissolved.

Penicillin wasn't just a drug. It was a turning point. The first true antibiotic, and the proof that pharma could conquer disease at scale.

The Legacy

Penicillin changed the world because it:

- Saved millions of lives during wartime and beyond
- Launched the modern antibiotic revolution
- Inspired a new generation of scientists and pharmaceutical companies
- Proved that accidents in science can lead to miracles
- Showed that urgency, collaboration, and belief could accelerate discovery

It was, and still is, one of humanity's greatest gifts—grown in mold, born of war, and transformed into hope.

And yes, penicillin won the war.

Not on the battlefield, but in the bloodstream.

Alexander Fleming, the Accidental Mold, and Global Salvation

How a Quiet Scientist and a Forgotten Petri Dish Changed the Fate of Humanity

In the summer of 1928, a quiet, unassuming scientist named Alexander Fleming returned from a two-week holiday to his perpetually cluttered lab at St. Mary's Hospital in London. He wasn't expecting to find anything extraordinary—just some old experiments, half-finished notes, and likely, bacterial cultures to toss in the bin.

But among the mess, he noticed something odd.

One of his petri dishes, filled with colonies of Staphylococcus aureus—a common but dangerous bacterium—had been invaded by a patch of blue-green mold. Around the mold, the bacteria had completely disappeared, forming a clear halo of inhibition.

Instead of throwing it away, Fleming leaned in.

"That's funny," he muttered.

"The bacteria don't seem to grow near this mold."

It was a simple observation.

It would become one of the most important discoveries in human history.

The Mold that Killed Bacteria

The mold was soon identified as belonging to the genus Penicillium. Fleming realized that it produced a substance that could kill a wide range of harmful bacteria—without harming the surrounding tissue.

He named this magical secretion "penicillin."

Over the next several months, he worked to extract and study it. He tested it against numerous bacterial infections—scarlet fever, diphtheria, pneumonia—and the results were consistent: the mold's extract killed them all.

He published his findings in a medical journal in 1929. But the paper received little attention. At the time, the medical world was skeptical of using natural substances as systemic treatments, and Fleming, being a bacteriologist rather than a chemist or clinician, struggled to purify and stabilize the compound.

And so, penicillin—the future savior of millions—sat quietly in obscurity for nearly a decade.

A Man of Modesty, a Mind of Precision

Fleming was not the sort to shout his discovery from the rooftops. Born to a Scottish farming family in 1881, he was methodical, introverted, and notoriously humble.

Colleagues often described him as "reserved," but his scientific instincts were sharp. He had already gained recognition for his earlier discovery of lysozyme, an enzyme found in human tears and saliva that killed some bacteria.

Yet Fleming never fully cracked the secret of penicillin's potential. His lab notes were careful but limited. He lacked the tools—and the funding—to take his discovery further.

Still, he preserved his petri dish, published his findings, and moved on. Unbeknownst to him, the seed had been planted.

Rediscovery and Resurrection

A decade later, as war loomed over Europe, two Oxford scientists—Howard Florey and Ernst Chain—stumbled upon Fleming's forgotten paper. They were actively searching for substances that could kill bacteria, and Fleming's observations sparked their curiosity.

Working with a small team (including Norman Heatley, who would engineer the crucial purification process), they finally managed to isolate and stabilize penicillin, making it suitable for human use.

Their first successful patient trials in 1941 confirmed what Fleming had always suspected: penicillin could treat bacterial infections that had previously been fatal.

Penicillin goes to War

As World War II intensified, the Allied forces realized the strategic importance of this new weapon. Penicillin was mass-produced and shipped to frontlines. Soldiers wounded in battle no longer died from infection. Post-surgical recovery soared. Amputations plummeted.

Penicillin saved more soldiers than any battlefield strategy ever could.

And while the industrial effort was driven by American pharmaceutical companies, the world never forgot who had first noticed the magic in the mold.

Recognition—But not Rewriting

In 1945, Fleming was awarded the Nobel Prize in Physiology or Medicine, alongside Florey and Chain. In typical modesty, he acknowledged that others had done the heavy lifting to develop penicillin into a usable drug. Yet his role as the first observer, the one who saw wonder in an accident, was rightfully honored.

His face graced magazine covers. He was knighted by the British Crown. He became a symbol of quiet genius—a man who saw what others ignored, and in doing so, gave the world its first true superdrug.

Global Salvation, Mold-Born

Today, penicillin and its descendants have saved hundreds of millions of lives. Entire generations were spared the scourge of diseases like syphilis, pneumonia, diphtheria, and gangrene because of that one moldy dish on Fleming's bench.

His discovery:

- Ignited the antibiotic era
- Reduced child mortality rates worldwide
- Extended human life expectancy by decades
- Inspired the rise of modern pharmaceutical research

And it all began with accident, curiosity, and an open mind.

The Lesson of Fleming

Alexander Fleming didn't invent penicillin.

He didn't refine it. He didn't commercialize it.

But what he did—see what others overlooked—was no less miraculous.

Sometimes salvation comes not with a bang, but with a halo around a mold.

And sometimes, all it takes to save the world is a scientist who's willing to look closer—and not clean up the lab too quickly.

Stories from the Frontline: WWII Soldiers Saved by Penicillin

How a Tiny Mold gave Life back to the Wounded—and Changed the Course of War

World War II was not just a war of bullets and bombs—it was a war of blood, infection, and survival.

The battlefield didn't just claim lives with shrapnel and gunfire. What came after the injury—the festering wound, the fever, the infection—was often far more lethal. Before penicillin, sepsis, gangrene, and pneumonia killed thousands of soldiers, even after they had been pulled to safety.

But by the time Allied troops stormed the beaches of Normandy in June 1944, there was a new weapon in their kits—not a gun, not a grenade, but a small vial of penicillin.

This was the first war in human history where medicine didn't just patch up the wounded—it brought them back from death's edge.

A Bullet Wound that didn't mean Goodbye

Private James Allen, a 23-year-old infantryman from Ohio, was hit in the thigh by machine-gun fire during the advance into Normandy. He was dragged into a field hospital, bleeding and broken. Medics treated the wound and stopped the bleeding, but the real fear was infection.

Without antibiotics, wounds like his often became septic within days. Legs were amputated. Lives were lost.

But James was among the first to receive penicillin intravenously. The drug stopped the bacteria in its tracks. Within 48 hours, his fever broke. The infection never came. His leg was saved—and so was he.

He later said, "I came to France with a rifle. I left with a limp—and my life. Thanks to penicillin."

D-Day's Silent Hero

On June 6, 1944, tens of thousands of Allied troops landed on the beaches of Normandy. The battle was fierce and bloody. Field medics worked around the clock. And for the first time in combat history, they had a new tool: syringes filled with penicillin.

In previous wars, such wounds would rot in muddy trenches. This time, medics injected soldiers within minutes of injury, halting infection before it started. It was like putting out a fire before it could spread.

According to Allied records, the number of amputations dropped by over 50% in field hospitals where penicillin was used.

Behind every saved limb was a soldier given another chance at life. Behind every survivor was a microbe stopped in its tracks by a miracle drug born of mold.

The Pilot who couldn't be Grounded

Flight Lieutenant David Walker, a Royal Air Force pilot, was wounded during a bombing run over occupied France. Shrapnel tore through his shoulder and lodged in his chest. Though stabilized, he developed bacterial pneumonia—the leading cause of death in war hospitals at the time.

Doctors administered a course of penicillin, and within days, the infection cleared. He went on to fly 35 more missions.

His survival wasn't just personal—it allowed his entire squadron to continue operating. "Penicillin," he later wrote in a letter home, "has done more for us than any aircraft design."

Medic's Best Friend: Penicillin in the Field Kit

For the first time, battlefield medicine had something other than gauze and morphine. Medics carried penicillin in specially designed kits—dry powder vials to be reconstituted with sterile water. They learned to inject it fast, often right in the field, when time was the difference between a saved limb and an amputation.

In Sicily, in the Pacific, in the forests of Belgium—wherever there were wounds, there was penicillin.

One medic, Corporal Raymond Ortiz, recalled:

"Before penicillin, you'd look at a wound and just know—this kid's not gonna make it. After penicillin, you looked and thought—he just might walk again. Maybe even dance."

The Greatest Weapon never Fired

By war's end, penicillin had treated over 2.5 million Allied soldiers. It saved lives without firing a shot. It turned wounds from death sentences into recoverable injuries.

Penicillin didn't win medals.

It didn't get monuments.

But in every saved soldier, in every grateful family back home, it became a silent hero of the war.

A Soldier, a Letter, and a Life Saved

One young Marine, shot in the Pacific theatre, wrote home after his recovery:

"They say I should be dead. That's what the docs told me. But they gave me this new medicine—they called it penicillin. I guess I owe that mold a hell of a lot."

And with that letter came a surge of national gratitude. Not just toward doctors and nurses, but toward the science that had risen to meet the horrors of war—and beat them back.

Legacy of the Frontline Lives

Penicillin's impact on WWII wasn't just clinical. It was cultural.

It proved that science could be a soldier, that molecules could march alongside men, and that a single accidental discovery could save millions of lives.

The war changed medicine. But penicillin changed war—by making survival possible on a scale never seen before.

Chapter 5

Streptomycin and the War on Tuberculosis

The Soil Bacteria, the Graduate Student, and the Cure that Saved Millions

In the early 20th century, tuberculosis (TB) was called "the white plague."

It wasn't just a disease—it was a slow, poetic death. It wasted the body, hollowed the chest, and stole breath from the lungs. People with TB were sent to sanatoriums in the mountains, where the only "cures" were rest, sunlight, and hope.

But hope rarely survived the night. In 1900, TB was the leading cause of death in America and Europe. It claimed millions every year—from factory workers to poets, mothers to kings.

And then, one day, in a modest laboratory in New Jersey, a clump of dirt changed everything.

The Dream of Selman Waksman

Ukrainian-born microbiologist Selman Waksman, a soil scientist at Rutgers University, believed the earth held secrets—not just for growing crops, but for healing bodies. He had spent decades studying actinomycetes, a group of bacteria that lived in soil and produced antibiotics to outcompete other microbes.

Waksman and his students collected thousands of soil samples from around the world, cultivating and testing them against harmful bacteria. His lab had already discovered several antibiotic compounds, but nothing that worked against TB.

Then, in 1943, a graduate student working under Waksman—Albert Schatz, only 23 years old—isolated a compound from a strain of Streptomyces griseus. When tested against TB bacteria, it killed them outright.

They named the compound streptomycin.

The First Real Weapon against TB

Unlike penicillin, which had little effect on TB, streptomycin could penetrate deep into infected tissues, especially the lungs. In early tests on guinea pigs, streptomycin eradicated tubercular infections. The results were so startling, the lab had to rerun them to believe them.

By 1944, streptomycin was being tested in humans.

One of the first patients was Patricia "Patsy" Jane Thomas, a young woman wasting away from TB in New York. After weeks of streptomycin injections, her fever vanished, her cough receded, and her strength returned. She became one of the first recorded TB patients to walk out of a hospital cured by a drug.

A World Rejoices

The impact was seismic.

In hospitals across the globe, patients once left to die slowly were now recovering. Sanatoriums, once filled with resigned sighs and whispered prayers, rang with hope.

Doctors watched with awe as X-rays showed lung cavities shrinking, coughs disappearing, and patients regaining weight. A death sentence had turned into a treatable disease.

The TB Epidemic Reversed

Streptomycin became the first truly effective drug against tuberculosis. It didn't work alone for long—resistance developed quickly—but it opened the door to combination therapy, the strategy that remains the backbone of TB treatment today.

- Isoniazid and para aminosalicylic acid (PAS) soon followed, and by the 1950s, doctors could treat TB with multi-drug regimens.
- The death rate from tuberculosis plummeted.
- Life expectancy soared.
- Sanatoriums emptied—some converted into hotels, others into museums.

Scientific Glory, Quiet Controversy

In 1952, Selman Waksman received the Nobel Prize in Physiology or Medicine for the discovery of streptomycin. But controversy brewed. Albert Schatz, the graduate student who had actually isolated the compound, felt overlooked. He eventually filed a lawsuit and was awarded co-credit in legal and historical records.

Waksman later admitted that Schatz had played a vital role but maintained that the discovery was part of the broader research vision he had led. History would go on to remember both their names—though not always equally.

The Legacy of Streptomycin

Streptomycin changed medicine in four profound ways:

1. It conquered the greatest infectious killer of the time.
2. It proved the power of systematic soil microbiology in drug discovery.
3. It introduced the concept of drug resistance and combination therapy.
4. It saved tens of millions of lives—and continues to do so today.

Even now, streptomycin remains part of the treatment regimen for multi-drug-resistant TB in some regions. Its discovery laid the foundation for dozens of other antibiotics derived from soil-dwelling microbes.

From a Handful of Soil, a Mountain of Hope

Out of a spoonful of dirt came a molecule that toppled a plague.

Out of a university lab came a cure for the poor, the sick, the voiceless.

Out of the hands of a student came a drug that turned death into life.

Streptomycin didn't just treat tuberculosis—it helped humanity breathe again.

Selman Waksman and the Soil Microbe that gave Hope

How a Humble Scientist, a Handful of Dirt, and a Hidden Microbe Defeated the White Plague

For centuries, tuberculosis (TB) was one of the deadliest diseases known to humanity.

It didn't come suddenly, like smallpox, or dramatically, like cholera. It came slowly—stealing breath, hollowing the chest, draining life over months and years. It took children, poets, mothers, workers, and artists. It was called "consumption" because it consumed the body from the inside out. In the early 20th century, one in seven deaths worldwide was due to TB.

Cures were few. Most relied on mountain air, fresh milk, and sanatoriums. But none of it worked.

And then one day, in a laboratory in New Jersey, a handful of soil gave the world something it had never had before: a cure.

The Man who Believed in Dirt

Selman Abraham Waksman was born in 1888 in a small village near Kyiv, in what is now Ukraine. He immigrated to the United States in his twenties and found work on a farm. There, he fell in love with the smell of earth, the texture of soil, and the idea that it was alive—not just dirt, but a universe of microbes.

He studied at Rutgers University and earned a Ph.D. in biochemistry. But Waksman wasn't interested in glamorous research. He believed the soil held secrets that could save lives—if we only knew how to listen.

And so, at Rutgers, he created the first structured soil microbiology program in the world.

The Idea that Would Change Everything

Waksman believed that microbes in the soil—particularly a group called actinomycetes—produced natural antibiotics to fight off other microbes. If these bacterial "defense chemicals" could be harnessed, they might help fight infections in humans too.

So he and his team did something no one had done at scale:

They systematically collected, cultured, and tested soil samples from across the country.

Fields. Gardens. Forests. Riverbanks. Even backyards.

Each sample was a potential gold mine. Each petri dish, a battlefield of microscopic war. They weren't just growing bacteria—they were searching for microbial warriors.

The Graduate Student who Found the Key

In 1943, one of Waksman's graduate students, Albert Schatz, isolated a strain of Streptomyces griseus from soil taken from a nearby farm. The compound it produced, they found, was effective not just against common bacteria—but against the tuberculosis bacillus, which had resisted penicillin and most known antibiotics.

They named the compound streptomycin.

And it worked. Spectacularly.

In guinea pigs, in test tubes, and soon, in people.

The Microbe that gave Hope

Streptomycin was the first antibiotic effective against tuberculosis. It turned sanatoriums from places of quiet decline into hubs of hope and healing. It gave breath back to the breathless. Weight returned to wasting bodies. Fevers broke. Death retreated.

In clinics around the world, doctors stood in awe as X-rays showed lung cavities shrinking. TB patients who had been written off got out of bed. Some left the hospital altogether—cured.

The microbe from a New Jersey soil sample had done what generations of healers could not.

Recognition and Controversy

In 1952, Waksman was awarded the Nobel Prize in Physiology or Medicine "for his discovery of streptomycin." He was lauded as the man who found the cure for TB—hailed by newspapers as a hero.

But behind the scenes, tension brewed. Albert Schatz, who had done the bench work and isolated the compound, felt overlooked and erased. Waksman, in turn, believed it was the vision and framework he created that made the discovery possible.

Eventually, Schatz sued for recognition. The case ended with Schatz being legally acknowledged as a co-discoverer of streptomycin and receiving a small share of royalties. But the Nobel remained with Waksman.

History has since remembered both men—Waksman as the architect of soil antibiotic discovery, and Schatz as the hands that uncovered one of medicine's greatest tools.

A Cure from the Earth itself

Selman Waksman once said:

"There is something infinite in the soil. Not just in quantity, but in possibility."

He was right.

Out of the soil came streptomycin. Out of streptomycin came the defeat of tuberculosis. Out of Waksman's vision came a revolution in pharmaceutical science—one that continues today in the search for new antibiotics from natural environments.

The war against TB is not over. But it was Waksman's soil, and that one tiny microbe, that first turned the tide.

TB Sanatoriums to Cures: Lives brought back from the Brink

How the Silence of Sanatoriums gave Way to the Sound of Recovery

For decades, tuberculosis was a slow-burning sentence.

It didn't steal life in a flash. It chipped away at it—breath by breath, cough by cough.

There was no cure. No drug. Only a ritual of retreat.

Patients were sent away to sanatoriums—remote institutions tucked in the mountains or deep forests, far from cities. There, the sick were prescribed rest, fresh air, milk, sunshine, and silence. It was less treatment and more containment. Hope was rationed. Recovery was rare.

By the early 20th century, hundreds of sanatoriums dotted Europe and North America. Entire economies and communities grew around these institutions, filled not just with patients, but with resignation.

Then came streptomycin. And with it, the greatest reversal in the history of medicine's fight against tuberculosis.

Life Inside the Sanatorium

The experience was almost monastic.

Patients lived in shared dormitories or solitary rooms, wrapped in blankets on outdoor porches even in the dead of winter, breathing what doctors hoped would be "clean" mountain air. They were forbidden from physical exertion, often instructed not to laugh, cry, or speak loudly.

There were reading hours. Writing hours. Resting hours. But mostly, there was waiting.

Waiting for the cough to ease.

Waiting for the fever to subside.

Waiting for breath to return.

Most never left.

Some would spend years in sanatoriums, dying slowly in surroundings filled with beauty but steeped in sorrow.

The First Streptomycin Trials: Hope on a Needle

In 1944, everything changed.

At the Mayo Clinic and other trial sites, streptomycin—a compound derived from Streptomyces griseus, a soil microbe—was administered to a small group of TB patients who were not responding to any traditional methods.

What followed was astonishing.

- Lung lesions began to heal.
- Weight returned.
- Persistent, hacking coughs stopped.
- Patients who had been too weak to sit up were walking.

Word of these recoveries spread fast. The halls of TB sanatoriums, once filled with muffled sobs and stillness, echoed with a new sound: laughter. Laughter from those who never expected to laugh again.

Stories of Lives Snatched from Death's Door

One patient, Helen Lawson, had been living in a Colorado sanatorium for three years. Her parents visited once a month, never knowing whether they'd be saying goodbye. She was pale, skeletal, and sinking. But after a few weeks on streptomycin, Helen gained weight, her fever dropped, and her X-rays showed improvement. A year later, she left the sanatorium, married her sweetheart, and lived into her 80s.

In London, a young man named Thomas Wainwright, who had worked as a factory laborer, had been bedridden in a ward for months. After starting streptomycin treatment, his cough vanished, and he returned to work within a year—the first in his ward to ever be discharged fully cured.

Across continents, such stories multiplied. The incurable were being cured.

The Sanatoriums Empty Out

As streptomycin became widely available, and other drugs like isoniazid and para-aminosalicylic acid (PAS) followed, TB treatment entered the era of combination therapy—more effective, more sustainable, and less prone to resistance.

Suddenly:

- Sanatoriums once overflowing with patients began closing their doors.
- Public health campaigns shifted from isolation to early detection and treatment.
- Families reunited.
- The stigma around TB faded.

By the 1960s, most Western countries had shuttered their TB sanatoriums. The haunting beauty of these institutions—once filled with whispered final goodbyes—stood eerily empty, now relics of a pre-antibiotic past.

From Isolation to Integration

Patients no longer had to disappear to recover.

With effective antibiotics, TB care was integrated into public hospitals, outpatient clinics, and home care. What was once a prolonged institutional sentence became a six-to-nine-month medical regimen.

The shift wasn't just medical—it was cultural. TB lost its romanticized "poet's death" reputation. It became treatable, beatable, survivable.

The Legacy of Survival

Millions of lives were brought back from the brink.

Children who once wasted away in mountain wards grew up, had families, became teachers, nurses, and writers.

Former patients became advocates, fighting for access to TB care in low-income countries.

And even today, as the fight against drug-resistant TB continues, the memory of those first cures fuels the mission:

No more dying in silence. No more years wasted in waiting.

A Revolution in White Coats and Soil

What streptomycin did wasn't just eliminate a disease—it liberated a generation.

From isolation.

From despair.

From slow decline.

It proved that the world could rise again, even from its most ancient and quiet killers.

And it all began with a soil microbe, a student's discovery, and a belief that even in the darkest corners of medicine, there is always something worth saving.

Chapter 6

Cortisone and the Fight against Inflammation

The Adrenal Hormone that gave the World its First Taste of Relief

In the 1940s, a 29-year-old woman named Henrietta Jackson could barely walk. Each morning, she woke with her fingers stiff and curled, her knees swollen like ripe plums, her spine locked in place. Diagnosed with rheumatoid arthritis, she had been told, like so many others, "Learn to live with the pain."

She did, until the day a doctor offered her an experimental injection—something newly extracted from deep inside the body's chemistry. Within hours, the swelling faded. By the next day, she could hold a pen. Within a week, she danced with her grandson.

The drug was cortisone.

And it marked the beginning of the steroid era—when inflammation, long seen as untouchable, could finally be controlled.

What is Inflammation, Really?

For centuries, physicians recognized inflammation by its signs: heat, redness, swelling, pain. But no one truly understood how to tame it.

Inflammation wasn't a single disease. It was the body's response to injury, infection, or internal confusion—as in autoimmune disorders like rheumatoid arthritis or lupus. It protected in the short term. But when chronic or misdirected, it became a slow fire that burned joints, organs, and lives.

And until cortisone, there was no way to extinguish that fire.

The Hormone Deep Within: Tracing the Source

In the 1930s, American biochemist Edward Kendall at the Mayo Clinic began isolating hormones from the adrenal cortex, a small part of the adrenal glands that sit atop the kidneys. The adrenal glands were known to affect metabolism and stress—but their internal workings were a mystery.

Kendall painstakingly extracted and named these compounds Compound A, Compound B, and so on—until he reached Compound E.

Later known as cortisone, Compound E had potent effects on carbohydrate metabolism—but also seemed to suppress inflammation. At the same time, German-American physiologist Philip Hench was observing something curious: arthritis patients often improved temporarily during pregnancy or after jaundice—times when hormone levels shifted dramatically.

Could something inside the body be responsible for taming the immune system?

The Cortisone Convergence

Kendall and Hench began working together, a rare marriage of chemistry and clinical insight. In 1948, they administered cortisone to patients with severe rheumatoid arthritis.

The results were astonishing.

Patients who had been bedridden for years stood up. Swollen joints returned to normal. Chronic pain melted away.

It was one of the most dramatic therapeutic turnarounds in medical history.

Henrietta Jackson, whose story opened this chapter, was one of hundreds who felt they had been resurrected.

From Miracle to Mainstay

In 1950, Kendall, Hench, and Swiss chemist Tadeus Reichstein (who had independently worked on adrenal hormones) were awarded the Nobel Prize in Physiology or Medicine.

Pharmaceutical companies raced to produce cortisone and its derivatives. Soon, hydrocortisone, prednisone, dexamethasone, and mcthylprednisolone followed—each more potent, more targeted, and more usable than the last.

Inflammatory diseases—once deemed incurable—now had a fighting chance:

- Rheumatoid arthritis
- Lupus
- Asthma
- Inflammatory bowel disease
- Severe allergic reactions
- Skin diseases like eczema and psoriasis

All became manageable with corticosteroids.

The Double-Edged Sword

But cortisone came with a price.

Long-term use led to serious side effects:

- Weight gain
- Osteoporosis
- Diabetes
- High blood pressure
- Mood changes
- Immune suppression

Patients once saved by cortisone began facing new health battles.

Doctors learned that corticosteroids weren't a panacea—they were a powerful but delicate tool, to be used with caution and respect. Over time, treatment shifted toward low doses, short courses, or localized delivery to reduce harm.

Still, when needed—when inflammation raged and life was at stake—cortisone remained unmatched.

A Turning Point in Immunology and Drug Discovery

Cortisone was more than a drug—it was a proof of concept.

It showed the world that:

- The immune system could be pharmacologically modulated
- Hormones could have medical applications far beyond their natural roles
- Diseases like arthritis weren't just fate—they were treatable

It also spurred the creation of an entire new class of medicines: anti-inflammatory drugs, both steroidal and non-steroidal (NSAIDs), shaping the pharmaceutical landscape for generations.

The Fire Within, Finally Tamed

For people like Henrietta Jackson, cortisone wasn't just a chemical.

It was a chance to reclaim the life they had lost.

It turned chronic illness from a slow descent into immobility into something manageable—livable—even invisible, at times.

And for medicine, cortisone marked the beginning of understanding inflammation not as a curse, but as a system that could be influenced, balanced, and, when necessary, calmed.

From Animal Glands to Synthetic Relief

How Pharma Transformed a Scarce Hormone into a Scalable Lifesaver

When cortisone first stunned the medical world in the late 1940s—helping arthritis patients stand, walk, and even dance again—it was hailed as a miracle drug.

But there was a catch.

The miracle was nearly impossible to mass-produce.

Early cortisone was derived from animal adrenal glands—tiny crescent-shaped organs that sit atop the kidneys. It took thousands of glands from cattle, painstakingly harvested and processed, to produce just a few milligrams of cortisone.

At the time, this meant that cortisone was rarer than gold—and more expensive than most patients could ever afford.

If it was going to help the world, the miracle needed a scalable solution.

The Quest for a Better Source

In the early 1950s, the pharmaceutical world scrambled to find alternatives. Chemists knew the structure of cortisone. They just needed a way to produce it synthetically or semi-synthetically, using more abundant precursors.

The search took them not to exotic animals or distant plants—but to an unlikely source:

Mexican yams.

Specifically, a wild yam species called Dioscorea mexicana, which grew in the jungles of southern Mexico. Its roots contained diosgenin, a steroid-like molecule that resembled the building blocks of cortisone.

The Rise of Steroid Chemistry: Russell Marker's Breakthrough

The key breakthrough came from an American chemist named Russell Marker, who, after working with sex hormones, realized he could extract diosgenin from yams and convert it into precursors of progesterone and cortisone.

He moved to Mexico, established Syntex, and launched what would become the steroid revolution.

Other pharmaceutical companies followed. Chemists refined the processes, learning how to build corticosteroids from these plant-based compounds. They no longer needed truckloads of adrenal glands. They had biochemistry, ingenuity, and roots.

From Jungle Roots to Pill Bottles

By the mid-1950s:

- Cortisone could be produced semi-synthetically at scale
- More potent and longer-lasting derivatives were developed (e.g., prednisone, dexamethasone)
- The drugs became affordable, available, and adaptable—in tablets, creams, injections, and eye drops

Pharmaceutical companies like Merck, Upjohn, and Pfizer mass-produced these new corticosteroids, placing them in pharmacies, hospitals, and emergency rooms around the world.

Doctors had a new tool to calm everything from raging arthritis to life-threatening asthma, from severe allergies to autoimmune flares.

And patients who once relied on hope alone now had relief in a bottle.

A Model for the Future

This journey—from animal extraction to plant-based synthesis—became a blueprint for pharmaceutical advancement:

- Start with nature
- Understand the structure
- Build better, safer, more accessible versions

It opened doors for synthetic versions of insulin, thyroid hormones, sex hormones, and eventually biologic drugs.

A Revolution Made Repeatable

The synthetic steroid revolution didn't just help patients—it changed how pharma thought:

- Diseases could be treated at their molecular roots
- Scarcity could be overcome by chemical ingenuity
- Nature could be honored—and improved upon

The relief that once cost a fortune and came in whispers now came in affordable, scalable doses, helping millions.

From adrenal glands to jungle yams to factory labs, cortisone's journey is one of the greatest testaments to the alchemy of science.

The Arthritis Patient who Walked again

One Woman's Journey from Crippling Pain to Unimaginable Relief— and the Miracle of Cortisone

In early 1949, a woman named Mrs. Gladys Roberts, 29 years old and once full of life, lay confined to a hospital bed at the Mayo Clinic in Rochester, Minnesota. Her hands were curled into claws. Her knees were locked in place. Every joint in her body felt as though it had been set on fire from the inside. She couldn't comb her hair, feed herself, or hold a pen.

She was one of thousands around the world suffering from severe rheumatoid arthritis, a condition where the immune system attacks its own joints, turning daily life into a cycle of pain, swelling, and slow deterioration. In the late 1940s, there was no real treatment. Only hot packs, splints, aspirin, and sympathy.

Gladys had lost her job, her independence, and nearly her will to live.

Then her doctors offered her a chance: an experimental compound known only as Compound E—later called cortisone. It had never been used this way before. No one knew what would happen.

The First Injection

On a cold winter morning, Gladys received her first injection.

There was no fanfare, just the quiet hope of a medical team watching history in the making.

That night, something changed.

Her fever dropped. The pain dulled. Her fingers, long stiffened into a gnarled pose, began to loosen ever so slightly. She slept for the first time in weeks.

By the second day, she could lift her hands off the bed.

By the third, she could swing her legs over the edge.

By the fifth, she stood—on her own.

On the seventh day, she walked down the hospital hallway, smiling.

Doctors Wept. Medicine Changed

The attending physician, Dr. Philip Hench, who had long suspected that hormones could affect inflammation, wrote in his notes:

"A patient once crippled… now ambulatory. Pain-free. Eyes bright. Cheeks flushed. A transformation I had never seen in medicine before."

It wasn't just a clinical success. It was a resurrection.

Word spread through the Mayo Clinic like electricity. Bedridden patients asked for "the injection." Doctors from around the country called in disbelief. What had begun as a hormone extracted from adrenal glands now looked like a gift from the gods.

And Gladys? She went home. She hugged her children again. She returned to her sewing. Her pain was not gone forever, but it had been conquered.

Why her Story Mattered

Gladys Roberts became the first public face of cortisone's potential—not a celebrity, not a scientist, but a housewife who could walk again. She represented millions quietly suffering from inflammatory diseases, now seeing a glimmer of life beyond pain.

Her case was published in medical journals and broadcast to the wider public, igniting global demand for cortisone—and launching an entire class of drugs we now call corticosteroids.

From Onc Woman to Millions

Today, cortisone and its descendants help people with:

- Rheumatoid arthritis
- Lupus
- Asthma
- Crohn's disease
- Eczema
- Severe allergic reactions

Every injection, every tablet, every cream owes its existence to those first trials—and to the courage of patients like Gladys, who agreed to try something unproven because they had nothing left to lose.

Walking, Again and Again

Gladys Roberts never sought fame. But her walk down that hospital hallway in 1949 became one of the most iconic early moments in modern pharmacology—a quiet walk that thundered through medical history.

She walked not just for herself, but for millions more who would follow.

Chapter 7

Polio Vaccine: From Fear to Freedom

How Science Ended a Childhood Nightmare and gave the World its Legs Back

In the summers of the 1940s and early 1950s, American parents lived in quiet terror.

Pools closed. Playgrounds emptied. Children were kept indoors behind drawn curtains. Not because of war, or poverty, or famine—but because of a virus that could paralyze a child overnight.

That virus was poliomyelitis—better known as polio.

It struck without warning, creeping silently through communities, often infecting without symptoms—until one day, a child couldn't lift a leg, or sit up, or breathe. Some would recover. Others would spend the rest of their lives in wheelchairs or leg braces. The most unfortunate would live trapped inside the iron lung, a metal tube that breathed for them.

In 1952, the U.S. alone saw nearly 60,000 polio cases, over 3,000 deaths, and tens of thousands of children permanently disabled.

And then, in a laboratory in Pittsburgh, a quiet, methodical doctor named Jonas Salk gave the world the answer it had been praying for.

Polio's Reign of Terror

Polio was a paradox: a disease that had existed for centuries but grew more feared as sanitation improved. As clean water systems reduced early-life exposure, children weren't developing natural immunity. So when outbreaks hit, they hit hard—especially in affluent, urban areas.

Franklin D. Roosevelt, America's president, was left paralyzed from the waist down by polio at age 39.

Entire towns would erupt in panic at a single case.

Parents whispered the word "polio" the way past generations had whispered "plague."

There were no drugs. No treatments. Only crutches, leg braces, rehabilitation centers—and hope.

The Reluctant Hero: Jonas Salk

Dr. Jonas Salk wasn't seeking fame. The son of Russian immigrants, he was driven by discipline, not charisma. But when given the opportunity to lead a polio vaccine project at the University of Pittsburgh, he poured himself into the task.

Instead of using a live, weakened virus like others were attempting, Salk took a radical but elegant approach:

He used a killed virus—chemically inactivated but still capable of triggering the body's immune defenses.

If it worked, it would be safer and easier to manufacture at scale.

Salk spent years testing and refining the vaccine, starting in monkeys, then moving cautiously to human volunteers—including himself, his lab team, and even his own children.

By 1954, the vaccine was ready for the largest medical trial in history.

A National Experiment in Courage

The 1954 field trial enrolled 1.8 million children—"Polio Pioneers," as they were called. It was a double-blind, placebo-controlled trial on an unprecedented scale, carried out in schools across the U.S.

On April 12, 1955, the results were announced to the world in a nationally broadcast event:

"The vaccine works. It is safe, effective, and potent."

Across the country, church bells rang. Parents wept. Teachers hugged students.

Salk became an instant national hero.

And the country exhaled—for the first time in years.

From Clinics to Communities

Within weeks, vaccine production ramped up. Children stood in long lines for their shots, sleeves rolled, bravery showing in their eyes. By the late 1950s:

- Polio cases in the U.S. dropped by more than 85%
- Similar campaigns began worldwide
- Polio wards began to empty
- The iron lungs fell silent

A disease that once paralyzed cities was now being paralyzed by science.

And in 1961, a second vaccine—Albert Sabin's oral polio vaccine—was introduced, making mass immunization even easier, especially in developing countries.

From Fear to Freedom

The emotional transformation was staggering.

Children played outside again.

Public swimming pools reopened.

Parents stopped holding their breath every summer.

A generation that had grown up fearing leg braces now grew up running, dancing, climbing trees—free.

For many survivors, the vaccine arrived too late. But they became the strongest advocates, determined to ensure no child would suffer as they had.

The Man who Refused to Patent the Cure

When asked who owned the patent for the vaccine, Salk famously replied:

"There is no patent. Could you patent the sun?"

He had no interest in profiting from the discovery. His focus was on eradicating polio from the face of the earth.

That dream is now nearly fulfilled. Thanks to the efforts of the Global Polio Eradication Initiative, 99.9% of polio cases worldwide have been eliminated.

The Legacy

The polio vaccine didn't just save lives. It:

Launched the modern era of mass vaccination

Proved the power of public-private cooperation

Gave medicine a new kind of heroism—quiet, methodical, and driven by compassion

From Crutches to Cartwheels

The generation that feared the iron lung now watches its grandchildren live in a world where polio is all but forgotten.

But it must never be erased from history—for the freedom we enjoy today was paid for by decades of fear, courage, and a shot in the arm that changed the world.

Jonas Salk's Crusade against Paralysis

The Relentless Pursuit of a Man who believed No Child should Live in Fear

To understand Jonas Salk, you must first understand the world he lived in—a world where children woke up unable to move their legs, where hospitals overflowed with rows of tiny bodies in iron lungs, where parents were told, "There's nothing we can do."

The 1940s and early 1950s in America were gripped by fear—not of war, but of polio. A virus so elusive, it struck silently and spared no one. One day a child would have a sniffle, the next day be unable to walk.

Paralysis was the great equalizer. It crippled rich and poor alike, including President Franklin D. Roosevelt, whose own battle with polio gave the virus a face.

To millions, the disease seemed unstoppable.

But to Jonas Salk, it seemed unacceptable.

A Scientist with Soul

Salk wasn't a typical researcher. Born in 1914 in New York City to poor immigrant parents, he was a first-generation American with a deep reverence for learning and a quiet defiance of limits. He trained not just in medicine, but in research—believing that medicine could not only treat but prevent.

While others accepted polio as a terrible fate, Salk saw it as a solvable problem.

He joined the University of Pittsburgh in 1947, taking a modest lab position in a converted women's dormitory. With limited resources but an unwavering focus, he began dissecting the virus and mapping its weaknesses.

A Bold, Unconventional Strategy

Many of his peers were pursuing a live-attenuated vaccine—one that used a weakened form of the virus. But Salk chose a different path: a killed-virus vaccine, inactivated with formaldehyde. This form couldn't replicate or revert to virulence, but it could still train the immune system to recognize and destroy the real virus.

It was safer. Simpler. But riskier, too. Some called it naive. Others called it reckless.

Salk didn't care about glory. He cared about results.

So he worked—day and night, with a small, fiercely loyal team. When he was ready to test the vaccine in humans, he didn't ask for volunteers. He stepped forward himself, injecting his wife and three sons, then himself.

If it didn't work, they would all be exposed.

It worked.

The Trial that Made History

In 1954, with cautious confidence, Salk launched the largest clinical trial in medical history:

- 1.8 million children across the U.S.
- Known as the "Polio Pioneers", these children were part of a double-blind, placebo-controlled study that would define modern vaccine testing.

On April 12, 1955, the results were announced to the world:

"Safe, effective, and potent."

America wept with joy. Church bells rang. Classrooms erupted in applause. Jonas Salk was hailed as a hero. But he refused to take personal credit.

When asked who owned the patent, he famously said:

"There is no patent. Could you patent the sun?"

He had just given the world one of the greatest medical gifts of all time—and asked for nothing in return.

More than a Vaccine — A Vision

Salk's crusade wasn't just about stopping polio. It was about reclaiming childhood, restoring public trust, and proving that science could liberate humanity from fear.

He later founded the Salk Institute for Biological Studies in La Jolla, California—not as a monument to his greatness, but as a haven for open, interdisciplinary science. There, he would continue to work on HIV vaccines and autoimmune disease, always guided by the belief that knowledge should serve humanity.

The Man who Refused to Rest

Until his death in 1995, Salk remained a tireless advocate for ethical science, public health, and the moral responsibility of medicine.

He never sought fortune.

He didn't build a pharmaceutical empire.

He built a legacy of selfless, people-first science.

The Crusade that Freed a Generation

Thanks to Jonas Salk, millions of children could run, swim, and climb without fear.

Thanks to Jonas Salk, paralysis no longer cast its shadow over every summer.

Thanks to Jonas Salk, the world now stands on the brink of eradicating polio forever.

His crusade was quiet, relentless, and braver than any battle fought with bullets.

It was a crusade of the mind.

Of the heart.

Of one man's belief that no child should ever lose their legs—or their life—to a virus that could be stopped.

"My Child Can Play Again"

Stories from the Polio Plague

When Fear Haunted Every Summer, and a Shot Gave Back More Than Health—It Gave Back Childhood

Before Jonas Salk's vaccine, summer didn't mean vacation.

It meant fear.

To parents across America, June didn't bring the promise of sunshine and freedom—it brought whispered warnings, shuttered swimming pools, and children kept indoors. Because that's when polio struck the hardest.

There was no pattern. One day your child had a sore throat. The next day, they couldn't move their legs.

And then came the diagnosis: polio.

There was no treatment. No medicine. Only braces, crutches, and—if the paralysis reached the lungs—an iron lung, where a child lay in a metal cylinder that breathed for them.

Some lived. Some didn't.

And for those who did survive, the price was often permanent: weakened limbs, twisted spines, lost dreams.

But then, in 1955, a voice announced across the radio:

"The vaccine is safe, effective, and potent."

And with that, a generation was rescued.

This chapter shares the stories of that generation—the ones who lived in fear, fought to survive, and whispered five words the world hadn't heard in years:

"My child can play again."

The Little Boy who couldn't Climb

In the summer of 1952, six-year-old Bobby Kramer from Chicago was the fastest runner on his block. Then one morning, he couldn't get out of bed. His legs felt like jelly. By nightfall, he was in a hospital, separated from his parents, his lower body paralyzed.

His mother, Ellen, sat outside the isolation ward every day for weeks, clutching a photo of him on his tricycle.

The doctors told her there was little they could do.

Three years later, Bobby received the Salk vaccine—part of a program for children who had already survived polio to prevent reinfection. That same summer, Ellen watched him stand, supported by a brace, and take three wobbly steps toward her.

"It was just three steps," she said, "but to me, it felt like watching him cross the finish line of a marathon."

The Iron Lung and the Birthday Balloon

Nine-year-old Susan Reynolds contracted bulbar polio in 1953. The virus attacked the muscles that controlled her breathing. She was placed in an iron lung. The machine hissed and clicked, expanding and compressing her chest mechanically.

Her mother decorated the inside of the machine's mirrored hood with balloons and drawings for her birthday.

When the vaccine was released, Susan's younger siblings received it immediately. They never got sick. Susan, who survived the ordeal, later became a nurse.

"If the vaccine had come one year earlier," her mother said, "Susan might've never needed that machine."

Playgrounds Reclaimed

By 1957, just two years after the vaccine's approval, polio cases in the U.S. dropped by nearly 60%. By 1961, they had fallen by more than 90%.

Parents no longer clutched their children at the sound of a cough.

Backyard birthday parties resumed.

Public pools, once seen as viral hotspots, overflowed with laughter again.

One Ohio mother wrote to her local newspaper:

"This year, I watched my daughter jump rope again. Not with fear, but with joy. Thank you, Dr. Salk. Thank you for letting my child be a child again."

A World without Crutches

In schools across America, leg braces were removed. Orthopedic wards were repurposed. For the first time in decades, physical therapists weren't treating trauma—they were preventing it.

Children who had once waited in long lines for iron lungs now stood in line—for vaccines.

And when they emerged, they weren't just protected.

They were free.

The Unseen Heroes

Teachers who carried paralyzed students to class.

Doctors who refused to abandon the worst cases.

Parents who lost children, yet stood in line to vaccinate their next.

Survivors who grew up to become advocates, scientists, therapists.

They all share one truth:

The vaccine didn't just stop a virus. It restored dignity, movement, and futre.

"My Child can Play Again"

Those words echoed across continents.

From rural Alabama to the slums of Calcutta, from suburban Los Angeles to war-torn Korea—wherever the vaccine reached, it carried with it not just immunity—but innocence restored.

It gave parents a chance to say goodbye to fear.

It gave children a chance to say hello to trees they could climb again.

To fields they could run across.

To a world where their bodies did not betray them.

The Promise Fulfilled

Today, polio remains endemic in only two countries. The rest of the world lives in the afterglow of one of medicine's greatest victories.

And in the hearts of millions of mothers and fathers, one sentence will always define that moment in history:

"My child can play again."

PART III: THE GOLDEN AGE OF DRUG DISCOVERY (1970s-1990s)

Part III

The Golden Age of Drug Discovery (1970s - 1990s)

When Molecules Met Momentum—And Medicine Leapt Forward

If the post-war years gave the world its first taste of what pharmaceutical science could do, the decades that followed—the 1970s through the 1990s—unleashed a tsunami of innovation that redefined what medicine meant.

This era, often called the Golden Age of Drug Discovery, was not built on one miracle drug or one lab, but on a thousand brilliant steps taken across chemistry, biology, and technology.

It was the time when pharmaceutical companies transformed from manufacturing houses into engines of discovery, when laboratories became launchpads, and when diseases that once had no names gained not only labels—but treatments.

This was when science stopped reacting to disease—and began to anticipate it.

The Foundations of the Boom

Several breakthroughs laid the groundwork for this golden age:

- Receptor theory and molecular pharmacology matured, allowing scientists to design drugs that precisely targeted disease pathways.
- High-throughput screening enabled researchers to test thousands of compounds rapidly.
- Advances in X-ray crystallography and NMR helped reveal the shape of target proteins—ushering in the era of structure-based drug design.
- Computational chemistry and the first applications of artificial intelligence (even in primitive form) began shaping early drug modeling.

And, for the first time, governments around the world invested heavily in regulatory science, building frameworks that encouraged both innovation and safety.

The Blockbusters Arrive

By the late 1970s and into the 1980s and 1990s, a wave of blockbuster drugs flooded the market—medicines that not only transformed lives, but reshaped entire industries.

Here are just a few of the most iconic breakthroughs from that era:

Beta Blockers and ACE Inhibitors – The Heart Tamed

- Propranolol, the first widely used beta blocker, revolutionized the treatment of high blood pressure and heart disease. It reduced heart rate, blood pressure, and saved countless lives after heart attacks.
- Captopril, the first ACE inhibitor approved in 1981, offered a new way to treat hypertension and congestive heart failure by acting directly on the renin-angiotensin system.

For the first time, patients at risk of heart attacks had daily protection in a pill.

Statins – The Cholesterol Revolution

- In 1987, lovastatin (Mevacor) became the first statin approved for cholesterol control.
- These drugs blocked the liver enzyme responsible for cholesterol production—dramatically lowering LDL and reducing the risk of heart disease and stroke.

By the end of the 1990s, statins were among the most prescribed drugs in the world, changing cardiology forever.

Prozac – The Chemistry of the Mind

In 1987, fluoxetine—better known as Prozac—became the first SSRI (Selective Serotonin Reuptake Inhibitor) approved for depression. It offered not just efficacy, but a better side effect profile than older antidepressants.

Prozac didn't just treat depression—it helped destigmatize it. It became a cultural icon, sparking conversation, controversy, and compassion.

Mental health was no longer something to hide. It was something to treat.

Omeprazole – Acid Under Control

Approved in 1989, omeprazole (Losec/Prilosec) was the first proton pump inhibitor, changing the management of ulcers, GERD, and acid-related diseases.

What once required surgery or long-term discomfort was now a manageable condition.

Tamoxifen – A Turning Point in Cancer

Discovered earlier but gaining prominence in the 1980s, tamoxifen became the first targeted hormonal therapy for estrogen receptor-positive breast cancer.

It gave hope to millions of women and proved that targeted cancer therapy could work. It set the stage for future breakthroughs like Herceptin and Gleevec.

AZT – Hope in the Time of HIV

In 1987, AZT (zidovudine) became the first drug approved to fight HIV/AIDS. Though imperfect, it marked a break in the darkness.

It wasn't a cure—but it was a beginning. The 1990s saw the rise of HAART (Highly Active Antiretroviral Therapy), turning HIV from a death sentence into a chronic, manageable disease.

Big Pharma Grows Up

During this period, pharmaceutical companies:

- Poured billions into R&D
- Globalized their operations
- Engaged in strategic mergers and acquisitions
- Professionalized clinical trials and regulatory affairs

The phrase "Big Pharma" was coined—not always with admiration—but undeniably reflecting the scale of impact these companies had achieved.

Some of the biggest names in medicine today—Pfizer, Merck, GlaxoSmithKline, Roche, Novartis, AstraZeneca—were shaped or born during this golden age.

Patients became Participants

This was also the era when patients 'voices grew louder:

- Advocacy groups pushed for faster approvals and more inclusive trials.

- Conditions like AIDS, breast cancer, Parkinson's, and mental illness gained visibility.
- Public health campaigns aligned with pharmaceutical science, fueling prevention and patient empowerment.

The Legacy of the Golden Age

By the end of the 1990s:

- Life expectancy had risen dramatically across the developed world.
- Many once-lethal diseases became treatable or manageable.
- A new generation of scientists, inspired by the power of pharma, entered the field.

But perhaps the greatest legacy was this:

For the first time in human history, millions of people took pills each day not to survive illness—but to prevent it.

The pharmaceutical industry had moved from the sickbed to the center of life itself.

Chapter 8

Beta-Blockers and Heart Disease

How a New Class of Drugs Took the Fear Out of Every Heartbeat

In the early 1960s, a heart attack was not just a medical emergency—it was a life sentence. If you were lucky enough to survive the initial attack, the fear never left you. Every skipped beat, every chest flutter, every climb up the stairs was haunted by one thought:

"Is this the next one?"

Doctors could offer little reassurance. Bed rest, lifestyle advice, maybe some nitroglycerin. But for millions of people with high blood pressure, angina, or post-heart attack trauma, the medical toolkit was dangerously thin.

Then came a class of drugs that changed the rhythm of medicine—beta blockers.

They didn't just slow the pulse.

They slowed the panic.

And for the first time, doctors could look their heart patients in the eye and say:

"We can manage this."

Understanding the Problem: A War inside the Chest

Heart disease was already the leading cause of death worldwide. It wasn't a single illness, but a complex orchestra of dysfunction—high blood pressure, narrowed arteries, overactive stress hormones, damaged heart tissue.

At the center of it all was a hyperactive nervous system. The body's "fight or flight" response—meant to help us survive danger—was, over time, killing people. Adrenaline surged. Heart rates spiked. Blood vessels tightened. And for those with cardiac vulnerability, this was a slow, internal war.

Doctors knew they had to calm the storm. But how?

Enter James Black—and the Receptor Revolution

In the 1950s, a brilliant Scottish pharmacologist named Dr. James Black had a radical idea: instead of blocking adrenaline's effects everywhere, what if we could block them only in the heart?

The human heart, it turns out, has beta-adrenergic receptors—proteins that respond to adrenaline by increasing heart rate and force of contraction. Black set out to block those receptors, effectively dampening the heart's overreaction to stress.

In 1964, he and his team developed propranolol, the world's first beta blocker.

It worked. Brilliantly.

For patients with angina (chest pain from narrowed coronary arteries), propranolol reduced the heart's workload, preventing attacks before they started.

For those with hypertension, it lowered blood pressure by calming the cardiovascular system.

And for survivors of heart attacks, it became a shield against future death.

A Pill that Calmed the Heart—and the Mind

Patients described it as a deep exhale after years of holding their breath.

A man in London who had lived with crushing angina for years said after starting propranolol:

"It was like someone turned down the volume in my chest. I could walk again. Breathe again."

Another patient, recovering from a near-fatal heart attack, said:

"I still feared dying. But for the first time, I didn't fear living."

From the Clinic to the Corner Pharmacy

By the 1970s and 1980s, beta blockers became standard treatment worldwide for:

- High blood pressure
- Angina
- Arrhythmias

- Post-heart attack survival
- Heart failure (selectively)
- Even performance anxiety and migraines

Drugs like atenolol, metoprolol, bisoprolol, and carvedilol followed propranolol—each one more refined, more targeted, and better tolerated.

The class grew so rapidly that by the 1990s, beta blockers were among the most prescribed drugs on Earth.

Saving Lives by the Millions

What impact did they have?

In clinical trials, post-heart attack mortality dropped by 25–30% with beta blockers.

Hospitalizations for angina and arrhythmias plummeted.

Blood pressure management became predictable, safer, and more widely accessible.

Cardiologists could now prevent—not just react to—heart disease progression.

The Nobel Prize and the Man behind the Molecule

In 1988, James Black received the Nobel Prize in Physiology or Medicine for his pioneering work—not just for beta blockers, but also for H2 blockers like cimetidine, which revolutionized ulcer treatment.

Quiet and fiercely principled, Black believed in targeted pharmacology—not just dulling symptoms, but understanding and interrupting specific disease mechanisms.

He didn't just help manage heart disease.

He helped reshape the entire discipline of rational drug design.

More than a Drug—A Turning Point

Beta blockers signaled a philosophical shift:

- From crisis management to chronic disease control
- From generalized treatments to precision targeting
- From passive care to proactive protection

They gave patients the ability to live normally with heart disease, rather than in fear of it.

A Steady Beat of Progress

Today, beta blockers remain essential. They're paired with ACE inhibitors, statins, diuretics, and anticoagulants in comprehensive heart care protocols.

They've become so integrated into medicine that it's easy to forget just how revolutionary they once were.

But every time someone survives a heart attack and gets to walk their daughter down the aisle, or every time a man with hypertension avoids a stroke because of a small white tablet…

…we're reminded of what they really are:

Pills that bought people time—and gave the heart a fighting chance.

James Black and the Battle against the Silent Killer

How a Quiet Pharmacologist Tamed the Chemistry of Fear and Saved Millions of Hearts

In the quiet corridors of medicine, there are few diseases more deceptive than hypertension—high blood pressure. It gives no warning. No early symptoms. No dramatic collapse. Just a silent, steady pressure that wears down the arteries, weakens the heart, and invites stroke, heart attack, and death.

For decades, doctors called it what it was:

"The silent killer."

And while the world's attention was often drawn to more visible health threats—infectious diseases, cancers, and epidemics—this one lurked everywhere, taking lives by the millions.

Then came a man who believed the key to defeating it wasn't more drugs or surgeries, but a deeper understanding of how the body works.

His name was Sir James Whyte Black.

And he didn't just help fight hypertension—he helped revolutionize how drugs are discovered.

A Pharmacologist, Not a Physician

Born in 1924 in Uddingston, Scotland, James Black was no typical drug hunter. He wasn't flashy. He didn't run large labs. He wasn't even trained as a doctor.

But he had something rarer:

A belief that you couldn't treat a disease until you understood it—chemically, structurally, biologically.

Early in his career, while studying physiology, he became fascinated by the fight-or-flight response—the cascade of adrenaline that readies the body for danger. He realized this same system, when overactive, could also cause damage, especially to the heart.

Black asked a question no one else was asking at the time:

"What if we could block the heart's response to adrenaline—but leave the rest of the body untouched?"

Inventing a New Way to Calm the Heart

At a time when hypertension and angina were managed with rest, sedation, or rudimentary medications, Black envisioned a new kind of drug—one that would selectively block beta-adrenergic receptors in the heart, slowing its rate and reducing its demand for oxygen.

Working at ICI Pharmaceuticals (now part of AstraZeneca), he and his small team developed propranolol, the first beta blocker, in 1964.

When administered to patients with high blood pressure or chest pain, the effects were astounding:

- Heart rate slowed
- Blood pressure dropped
- Chest pain stopped
- Sudden death from heart attacks declined

It was the first drug designed to target a receptor—a specific molecular lock with a tailored key.

Changing Medicine Forever

Propranolol didn't just help people breathe easier—it changed their future.

A teacher once too weak to climb stairs returned to her classroom.

A businessman who feared collapse during meetings reclaimed his confidence.

An elderly man who had survived a heart attack now lived—not just longer, but better.

Beta blockers, refined and extended into drugs like atenolol, metoprolol, and bisoprolol, became cornerstones of cardiology. They helped manage:

- Hypertension
- Angina
- Arrhythmias
- Heart failure
- Even anxiety, migraines, and thyroid storm

They brought science closer to the precision medicine we know today.

The Quiet Genius behind it All

James Black never sought the spotlight.

He wasn't interested in commercial success.

He preferred quiet labs, clear thinking, and colleagues who valued questions more than fame.

He later joined SmithKline & French, where he helped develop cimetidine—the first H2-receptor antagonist for peptic ulcers. Another blockbuster. Another paradigm shift.

In 1988, he was awarded the Nobel Prize in Physiology or Medicine, not just for his discoveries, but for his approach—the idea that rational drug design could replace trial-and-error pharmacology.

"The most fruitful basis for the discovery of a new drug," he once said, "is to start with a deep understanding of how the body functions in health and disease."

Legacy: Taming the Silent Killer

Today, hypertension still affects over a billion people worldwide. But millions are alive—working, walking, parenting, thriving—because of the drugs James Black made possible.

- Every pill that slows a heart rate.
- Every quiet breath taken without chest pain.
- Every life not lost to a sudden stroke or cardiac arrest.

All trace back to a man who looked at adrenaline—not with fear, but with curiosity—and asked,

"What if we could make the heart calm down?"

He fought the silent killer not with drama, but with precision.

Not with noise, but with clarity.

And in doing so, James Black gave the world not just more time—but more peace.

Millions Saved from Sudden Cardiac Death

How Modern Medicine Turned Heart Attacks from a Death Sentence into a Detour—and gave the World its Fathers, Mothers, and Grandparents Back

There was a time—just a few decades ago—when heart attacks arrived like lightning: sudden, brutal, and often final. A man could be laughing at dinner, then slump in his chair. A woman could be shopping and collapse in the aisle. Many died before they reached the hospital, and those who survived the first attack often lived in fear of the second.

These were not slow deaths.

They were stolen moments, tearing loved ones away without warning.

The world called it "sudden cardiac death."

Medicine called it the hardest problem in cardiology.

But beginning in the 1970s and accelerating through the 1980s and 1990s, something extraordinary happened:

We began to turn the tide.

And millions of lives were quietly, powerfully saved.

Understanding the Enemy: When the Heart goes Rogue

Sudden cardiac death (SCD) is typically caused by ventricular fibrillation, a chaotic, disorganized rhythm in which the heart quivers instead of pumping. It can follow a heart attack, a heart failure episode, or appear without warning in seemingly healthy individuals.

Once it starts, the window to save a life is measured in minutes.

For decades, we were helpless.

Until medicine found its weapons—and used them with stunning success.

Beta Blockers: Smoothing the Storm

The first breakthrough came with beta blockers—medications like propranolol and metoprolol that calm the heart, slow it down, and reduce its response to stress hormones like adrenaline.

Clinical trials in the 1980s showed that beta blockers, given after a heart attack, could reduce the risk of sudden cardiac death by up to 30%.

That meant:

- Fathers lived to walk their daughters down the aisle.
- Mothers watched their children graduate.
- Grandparents got the chance to tell bedtime stories—and live long enough to hear them told back.

ACE Inhibitors and Statins: Building a Stronger Heart

Next came ACE inhibitors (like enalapril and lisinopril), which lowered blood pressure and remodeled damaged hearts, and statins, which dramatically lowered cholesterol and stabilized vulnerable plaques in the arteries.

These weren't just chemical fixes.

They were lifelines, preventing the heart from reaching that fatal tipping point.

For the first time, patients didn't just survive their first heart attack—they thrived beyond it.

A Revolution in the Ambulance: Defibrillators and Rapid Response

While medications transformed long-term survival, defibrillators transformed the moment of crisis itself.

Portable external defibrillators (AEDs) began appearing in malls, airports, and stadiums.

Emergency medical services were trained to recognize and treat SCD on-site.

Time-to-shock was reduced from minutes to seconds—and survival rates soared.

Today, more than 350,000 Americans suffer out-of-hospital cardiac arrest every year. Thanks to defibrillators and trained responders, tens of thousands are saved annually.

The Implanted Guardian Angel: ICDs

In the 1980s, the next evolution arrived: the Implantable Cardioverter-Defibrillator (ICD)—a matchbox-sized device placed under the skin, wired directly into the heart.

If a fatal rhythm began, the ICD would detect it and shock the heart back into rhythm—within seconds.

It was like carrying a guardian angel inside your chest.

Millions of high-risk patients have since received ICDs. Many owe their lives to jolts they never saw coming—but that brought them back from the brink.

Real Lives. Real Miracles.

A retired teacher in Michigan collapsed in her garden. Her ICD fired—twice. She woke up groggy, bruised, but alive.

A young athlete with a rare genetic heart condition was warned she might not live past 20. With an ICD and medication, she ran a 10K at 30.

A grandfather in Spain suffered a cardiac arrest at an airport. An AED shocked him back. He later held his newborn granddaughter and whispered, "Thank you for the second chance."

The Numbers Tell the Story

Since the 1980s:

- The sudden death rate after heart attacks has been cut in half in many countries.
- More than 7 million people are alive today thanks to beta blockers, ACE inhibitors, statins, ICDs, and emergency defibrillation.
- Hospital cardiac care units now routinely discharge survivors who, in earlier eras, would have never made it through the night.

More than Lives Saved —Futures Restored

These aren't just statistics. They're weddings. Graduations. Sunday dinners. Laughter. Retirements. Books written. Songs sung. First steps watched. Final hugs delayed.

There are millions of second chances, built by science—and made real by compassion.

From Sudden Death to Sustainable Life

Sudden cardiac death is still with us. But it's no longer a mystery.

No longer a sentence.

No longer a certainty.

Thanks to decades of relentless innovation, it is now—more often than not—a crisis we can beat.

And that's how medicine moved the needle from loss to legacy, from heartbreak to heartbeat.

Chapter 9

SSRIs: Giving Hope to the Depressed

The Daylight that Returned after Decades of Darkness

For centuries, depression was described with metaphors:

A black dog.

A fog that wouldn't lift.

A numbness that wrapped around the soul.

And for centuries, medicine struggled to help. The condition was misunderstood, stigmatized, even moralized. Sufferers were told to "snap out of it" or "count their blessings." What most couldn't see was that depression wasn't sadness—it was a neurochemical prison, invisible on the outside and unbearable on the inside.

Then, in the late 1980s, the world met a new class of medicines—ones that didn't sedate or blunt or shock, but gently, gradually, lift the veil.

They were called SSRIs—Selective Serotonin Reuptake Inhibitors.

And for millions, they were the first real source of hope.

Before the Breakthrough: Treatments that often Hurt

Before SSRIs, depression was treated with:

- Tricyclic antidepressants (TCAs) — effective but laden with side effects like dry mouth, weight gain, dizziness, and cardiac risk.
- MAO inhibitors (MAOIs) — powerful but dangerous when taken with certain foods or medications.
- In severe cases, electroconvulsive therapy (ECT) or institutionalization.

These tools, while life-saving for some, often intimidated or scared patients away from seeking treatment at all. Many suffered in silence.

Doctors knew they needed something gentler, something that wouldn't trade pain for numbness.

The Serotonin Hypothesis

In the 1960s and '70s, researchers began to suspect that low levels of serotonin—a neurotransmitter involved in mood regulation—might play a key role in depression.

The theory wasn't perfect, but it opened the door to a new idea:

What if we could selectively boost serotonin in the brain without triggering a host of other side effects?

This became the guiding light for the development of SSRIs.

The Arrival of Prozac: A Cultural and Medical Earthquake

In 1987, the FDA approved fluoxetine—better known by its brand name: Prozac.

It was the first SSRI.

Unlike previous drugs, it didn't cause sedation, weight gain, or dangerous interactions. It could be taken once daily, and patients often felt better within weeks.

Prozac didn't just work. It redefined what treatment looked like.

- Therapists reported breakthroughs in previously unreachable patients.
- Primary care physicians began prescribing it widely.
- Patients who had hidden their depression for years came forward.

For the first time, people said aloud:

"I'm on medication for depression."

And for the first time, it wasn't whispered.

Stories of Transformation

A teacher who hadn't left her house in a year returned to the classroom.

A father who'd withdrawn from his children said he finally felt "present" again.

A teenager, on the brink of ending his life, described the medication as "putting a floor beneath me so I didn't keep falling."

These weren't just chemical changes. They were restorations of identity.

A Class Expands, and Millions Heal

After Prozac, more SSRIs followed:

- Sertraline (Zoloft)
- Paroxetine (Paxil)
- Citalopram (Celexa)
- Escitalopram (Lexapro)
- Fluvoxamine (Luvox)

Each brought subtle differences—faster onset, fewer side effects, different receptor profiles—but they all shared the same core mechanism:

Increasing serotonin availability in the brain.

Today, SSRIs are prescribed for:

- Major depressive disorder
- Generalized anxiety disorder
- Panic disorder
- PTSD
- OCD
- Premenstrual dysphoric disorder

And more.

From Shame to Treatment

Perhaps the most powerful impact of SSRIs wasn't just medical—it was cultural.

They helped:

- Normalize conversations about mental health
- Encourage millions to seek help who might never have otherwise
- Shift depression from being a personal failing to a treatable condition

Books like Listening to Prozac sparked debate, yes—but they also spread understanding. The idea that "you don't have to feel this way forever" reached living rooms, classrooms, workplaces.

Not a Cure-All, but a Catalyst

SSRIs aren't perfect.

They don't work for everyone.

Some cause side effects.

Some patients need additional therapy or different medications.

But for many, they were the first stepping stone out of despair.

They gave space for therapy to work. For relationships to rebuild. For self-worth to re-emerge.

The Light that Returned

Since their introduction, over 300 million people worldwide have taken SSRIs. Countless lives have been changed—not just extended, but enriched.

In every city, in every family, someone once trapped in silence found their voice again because a tiny pill helped their brain hear it.

SSRIs didn't just treat depression.

They gave people back to themselves.

The Prozac Story and the Rise of Modern Psychiatry

The Little Green-and-White Pill that Sparked a Global Conversation—and a Medical Revolution

In the early 1980s, psychiatry stood at a crossroads.

Mental illness was real—but often hidden. Depression was common—but rarely named. And psychiatry, still haunted by the legacies of asylums, Freudian couches, and electroshock therapies, was viewed by many with suspicion, even disdain.

There were medications, but they were blunt instruments.

They worked for some. But for many, the side effects were unbearable.

People feared being "zombified." They feared addiction.

And so millions—perhaps tens of millions—suffered in silence.

Then came a pill that didn't look like salvation.

Small. Green and white. Taken once a day.

Its name was Prozac.

Its molecule was fluoxetine.

And what it unleashed was far more than a treatment.

It was the rebirth of modern psychiatry.

The Birth of a Breakthrough

In the late 1970s, at Eli Lilly & Co., a team of chemists and neuroscientists were searching for a new kind of antidepressant—one that would target specific brain chemistry without affecting every system in the body.

The working theory was this: in people with depression, the brain's supply of serotonin—a neurotransmitter involved in mood, sleep, and emotion—was too low. Older drugs like tricyclics affected many neurotransmitters and caused severe side effects. The goal was to increase serotonin levels—and only serotonin.

Enter fluoxetine—a compound discovered in 1974 and refined over several years. It worked by blocking the reabsorption (reuptake) of serotonin in the brain, keeping it active longer.

The term was born:

Selective Serotonin Reuptake Inhibitor—or SSRI.

1987: The Year the Conversation Changed

After nearly a decade of development and trials, Prozac was approved by the FDA in December 1987.

Its arrival was quiet. But its impact? Tectonic.

Within months, patients reported improvements without sedation, weight gain, or mental fog. They felt clearer. Calmer. Capable. They didn't feel "drugged"—they felt more like themselves.

By the early 1990s, Prozac had become a household name.

Doctors were prescribing it not just for major depression, but also for anxiety, OCD, panic disorder, and PTSD.

Patients—especially those who had resisted treatment out of fear or shame—were now asking for it by name.

It was the first time an antidepressant had become part of popular culture.

"Listening to Prozac" and the Cultural Reckoning

In 1993, psychiatrist Dr. Peter Kramer published Listening to Prozac, a book that asked provocative questions:

- If Prozac can make a shy person outgoing, are we treating illness—or enhancing personality?
- What is the line between treatment and transformation?
- If brain chemistry shapes who we are, do medications rewrite identity?

The book became a sensation.

Suddenly, antidepressants weren't just medical tools—they were philosophical lightning rods. Was Prozac changing people—or revealing their truest selves? Was psychiatry restoring health—or shaping humanity?

The debate was fierce. But the impact was undeniable.

The Face of Psychiatry Changes

Prozac forced the world—and the medical establishment—to take mental health seriously:

- Primary care doctors began treating depression routinely.
- Therapists and psychiatrists worked in tandem with medications.
- Patients gained agency, often choosing treatment paths with eyes wide open.

Psychiatry, once viewed as fringe, was now firmly rooted in neurobiology. The brain was no longer a black box—it was a system to be studied, mapped, and healed.

And Prozac was the door through which this transformation walked.

The Numbers and the Narratives

By the late 1990s:

- Prozac was prescribed to over 40 million people globally.
- SSRIs became the first-line treatment for depression worldwide.
- Suicide rates in several countries began to decline—especially among treated groups.
- Depression moved from taboo to topic of conversation.

Behind every statistic was a story:

- A teacher who returned to the classroom.
- A mother who laughed with her child again.
- A young adult who didn't end their life because a pill gave them enough relief to hope again

The Legacy—and the Lessons

Prozac wasn't perfect.

Some patients didn't respond.

Others experienced emotional flattening, sexual side effects, or discontinuation symptoms.

Critics warned of overprescribing.

Some feared psychiatry was becoming too chemical.

But one thing was clear:

Prozac changed the world's relationship with mental health—from silence to science.

It helped birth a generation of open conversations, early interventions, and evidence-based psychiatry.

It told millions who once felt broken:

"You're not weak. You're not alone. And you can get better."

More than a Pill—A Turning Point

Prozac was never just about serotonin.

It was about hope.

About believing that darkness could lift.

About giving people tools—not to erase their struggles, but to endure them and heal.

It sparked the rise of modern psychiatry, rooted in biology, informed by compassion, and driven by the belief that mental health deserves the same care and science as any other illness.

And for that, a little green-and-white pill will forever stand as a symbol of medicine's most powerful promise:

That healing is possible—even in the mind.

Stories of Silent Suffering Transformed into Living

When Chemistry Met Compassion—and Darkness Turned to Daylight

Depression rarely screams.

It doesn't show up with fevers or scans.

It whispers—quietly, persistently—until a person no longer recognizes their own reflection.

For decades, people with depression suffered in silence. They raised families, held jobs, passed strangers in grocery aisles—all while feeling like they were barely there. The world saw them functioning. What the world didn't see was the daily battle to get out of bed, the practiced smile, the exhaustion from pretending.

Then came a shift. The arrival of SSRIs and the growing awareness that depression was not weakness—but illness. That it wasn't shameful to feel broken—and it was no longer impossible to heal.

Here are their stories. Not of celebrities or headlines, but of ordinary people who quietly came back to life.

The Teacher who couldn't Stand in Front of her Class

Linda, a 34-year-old elementary school teacher in Seattle, had always been vibrant. But one winter, she started feeling detached— like she was watching her life from behind glass.

"I'd stand at the front of my classroom and forget why I was there," she said. "I couldn't feel anything. Not even fear."

Eventually, she stopped going. Her doctor diagnosed major depressive disorder and offered her fluoxetine—Prozac.

"I didn't believe a pill could help," she said. "But I was out of options."

Within weeks, she started to feel a flicker of energy. Then one day, a student from her old class sent her a card: "We miss your laugh."

That afternoon, she laughed—really laughed—for the first time in six months.

The Father who couldn't Feel Joy

Raymond, 52, was a father of three, a steady accountant, and—by all appearances—"fine."

But inside, he was hollow.

He loved his kids. He knew he did. But he hadn't felt it in years. Birthdays, holidays, milestones—he was there physically, but emotionally absent. He thought he was broken beyond repair.

When he finally spoke to his doctor, he was prescribed sertraline.

Three months later, he found himself at a Little League game. His son hit a double. The parents cheered. And Raymond felt a surge in his chest—not panic, not numbness. Pride.

"I didn't cry because I was sad," he said. "I cried because I realized I wasn't numb anymore."

The Teen who thought she'd always be this Way

Aisha, 19, was in college when the depression hit hard. She stopped attending lectures. Stopped eating. She would lie in bed staring at the ceiling, feeling nothing and everything at once.

Her roommate helped her get to counseling. Her psychiatrist prescribed escitalopram.

"The first thing I noticed," she said, "was that I started brushing my teeth again. Then I opened the curtains. Then one day, I found myself humming."

She didn't become a different person. She became herself again—the version of her not eclipsed by pain.

The Mother who thought she was Failing

Carla, 29, had just given birth to her second child. But instead of joy, she felt overwhelmed by sadness. She couldn't bond with her baby. She cried constantly. She thought she was a bad mother.

Her doctor recognized the signs of postpartum depression. With support, therapy, and medication, she began to reclaim her sense of self.

"The day my baby smiled at me—and I finally smiled back—that was the day I knew I was healing."

Millions of Lives, Quietly Reclaimed

These stories aren't rare. Since the late 1980s, hundreds of millions of people have taken SSRIs. Some needed them short-term. Others, long-term. Some paired them with therapy. Some used them as a bridge back to living.

The outcomes weren't always perfect. But for many, they were life-saving.

SSRIs gave people the strength to ask for help.

They gave doctors the tools to offer it.

And they gave families their loved ones back.

From Existing to Living

What these stories reveal is simple, yet profound:

- Depression is not a character flaw.
- Medication is not a moral failure.
- Healing is not only possible—it's worth fighting for.

Because sometimes, the greatest transformation is not dramatic.

It's quiet.

It's the moment someone says,

"I want to live again."

And means it.

Chapter 10

Statins: The Cholesterol Revolution

How a Simple Pill Changed the Fate of Millions of Hearts

For much of the 20th century, heart attacks struck like thunderbolts—sudden, fatal, and tragically common. They didn't discriminate. They ended the lives of truck drivers, teachers, world leaders, and athletes alike. Coronary heart disease was the leading cause of death worldwide.

And yet, for decades, there was a frustrating mystery at the heart of the crisis: cholesterol.

Doctors had long suspected a link between cholesterol and heart disease. Autopsies showed waxy plaques clogging arteries. But even as the evidence mounted, the question loomed:

Could we lower cholesterol—and if we did, would it save lives?

The answer came in the form of a pill that would become one of the most prescribed drugs in history. A class of medicines that not only lowered cholesterol but also redefined the prevention of heart disease.

They were called statins.

And they changed everything.

The Discovery in the Mold

In the 1970s, a Japanese biochemist named Dr. Akira Endo was searching for natural substances that could block HMG-CoA reductase—an enzyme in the liver that plays a key role in cholesterol production.

His inspiration? Penicillin—another breakthrough derived from mold. If antibiotics could fight bacteria, could fungi also contain agents that fought cholesterol?

From a strain of Penicillium citrinum, Endo isolated a compound he called compactin. It worked. It dramatically lowered cholesterol in animals.

Later, American pharmaceutical companies built on his work and developed lovastatin, the first statin approved for human use in 1987 by Merck.

It was a quiet revolution. But a powerful one.

The Proof Arrives: Saving Hearts, Not just Numbers

Skeptics wondered: Could lowering cholesterol on a blood test really prevent heart attacks?

In the Scandinavian Simvastatin Survival Study (4S) and later trials like WOSCOPS and HPS, the results were unequivocal:

- Statins reduced heart attacks and strokes by up to 40–50%.
- They cut cardiovascular death.
- They prolonged life, especially in those with previous heart events or diabetes.

Statins didn't just shift numbers. They saved lives.

How Statins Work

Statins inhibit HMG-CoA reductase, the enzyme responsible for producing cholesterol in the liver. By doing so, they lower:

- Total cholesterol
- Low-density lipoprotein (LDL) — the so-called "bad cholesterol"
- Inflammation in arterial walls, reducing plaque instability
- They also stabilize existing plaques, making heart attacks less likely.

In essence, statins turned cardiovascular care from a game of catch-up into one of prevention.

Real Lives, Real Impact

A 55-year-old man with high cholesterol and a family history of heart disease took simvastatin for ten years—and never had the heart attack that claimed his father's life at 48.

A woman with diabetes and rising lipid levels stayed healthy enough to see her grandchildren graduate.

A firefighter, told he was a "walking time bomb" due to inherited high cholesterol, began statin therapy at 40. At 65, he's still running drills—and still alive.

These aren't anecdotes. They are the quiet victories of modern medicine.

The Pill that Went Global

Following lovastatin came a new generation:

- Simvastatin (Zocor)
- Atorvastatin (Lipitor)
- Rosuvastatin (Crestor)
- Pravastatin, fluvastatin, and others

Lipitor became the best-selling drug of all time, with over $150 billion in global sales. But its real legacy isn't financial—it's epidemiological. It helped reshape the cardiovascular landscape.

- In countries with widespread statin use:
- Cardiovascular death rates plummeted
- Life expectancy rose
- Heart attacks became less fatal—and less frequent

From Reaction to Prevention

Statins allowed medicine to intervene before disaster struck. For the first time, doctors could look at a patient with elevated cholesterol and say:

"We can do something about this—starting today."

They also ushered in a preventive mindset:

- Screenings became routine
- Lifestyle + medication became the new standard
- Risk calculators replaced guesswork

This wasn't just medical progress—it was philosophical.

The Controversies and the Clarity

As statins became more widely prescribed, some pushed back:

- Were we "medicalizing" aging?
- Did side effects like muscle pain and fatigue outweigh the benefits?
- Were we too quick to medicate lifestyle problems?

These were fair questions. But across study after study, the data remained clear:

For those at moderate to high cardiovascular risk, statins dramatically reduce the chance of dying from heart disease.

The consensus?

Not everyone needs a statin.

But for those who do, it's one of the most effective preventive tools in medicine.

The Legacy of Statins

Today, statins are taken by over 200 million people worldwide.

They've:

- Prevented millions of heart attacks and strokes
- Bought decades of extra life
- Reduced fear in patients with family histories of heart disease

For many, they transformed cardiovascular risk from a ticking clock to a manageable condition.

They helped rewrite the future of aging.

A Revolution in a Pill

Statins weren't flashy. They didn't promise a cure.

But they did something more profound:

They quietly, steadily, saved lives—millions of them.

They gave families more dinners, more birthdays, more time.

They turned one of the world's most common killers into a condition that could be controlled.

And in doing so, they reminded us that sometimes the most powerful medicine is the one you take each morning without a second thought—because it lets you keep living the life you love.

■

The Curious Case of Mold and Lipitor

How a Fungus, a Forgotten Sample, and Relentless Curiosity Gave Birth to the Best-Selling Drug in History

When most people think of fungi, they think of rot, rash, or perhaps the mushrooms in their kitchen.

But time and again, mold has proven to be medicine's most surprising ally—from penicillin, which tamed infections, to statins, which tamed cholesterol.

The story of Lipitor, the best-selling drug in pharmaceutical history, begins not in a glamorous laboratory, but with a quiet mold sample, a discarded compound—and a chemist who refused to quit.

The Spark: Cholesterol's Infamous Reputation

By the 1970s, scientists had firmly established the link between elevated LDL cholesterol and coronary heart disease. The challenge wasn't understanding the problem—it was lowering cholesterol effectively and safely.

The race was on. Pharmaceutical companies around the world were investing heavily in cholesterol-lowering drug research. And it all led to one key enzyme:

HMG-CoA reductase—the gatekeeper of cholesterol synthesis in the liver.

Block this enzyme, researchers believed, and you block cholesterol production at its source.

Japan's Mold Discovery—and the Rise of Statins

In 1976, Japanese biochemist Akira Endo, working at Sankyo, isolated a compound from a strain of Penicillium citrinum—a mold found on rice.

He called it compactin. It inhibited HMG-CoA reductase in animals and lowered cholesterol. It was the first statin.

But compactin was never commercialized due to toxicity concerns in dogs. Still, it ignited a wave of development. Merck soon launched lovastatin, followed by simvastatin and pravastatin. These became standard therapies for high cholesterol.

It seemed the field was saturated.

But in a small lab at Warner-Lambert, a team led by chemist Bruce Roth believed something better was possible.

Lipitor Begins as a Long Shot

In the early 1980s, Roth's team synthesized a compound called atorvastatin. It was a thousand times more potent in vitro than lovastatin. But Warner-Lambert had already decided it was too late to enter the statin market. Merck dominated, and Lipitor—then just known as CI-981—was shelved.

But Roth couldn't let it go.

He pushed for further testing. Clinical results came back: atorvastatin consistently lowered LDL cholesterol more than any other statin—even at lower doses.

Warner-Lambert reversed its position and launched clinical trials. When the data came in, it was undeniable:

Lipitor wasn't just good. It was game-changing.

Pfizer Enters—and Lipitor Explodes

Recognizing its potential, Pfizer partnered with Warner-Lambert to co-market the drug. In 1996, Lipitor (atorvastatin) was approved by the FDA.

Pfizer took the unusual step of positioning Lipitor as a "superstatin," emphasizing its superior LDL-lowering ability and aggressive branding to physicians and patients.

By 2000, Lipitor:

- Became the world's most prescribed statin
- Was used by tens of millions globally
- Brought in over $12 billion annually at its peak

All from a molecule nearly forgotten, inspired by a mold growing quietly on rice in a petri dish.

The Mold behind the Miracle

Like penicillin before it, Lipitor traced its roots to fungal intelligence—the natural arms race between microbes, where compounds are created to outcompete other organisms. These microbial compounds—when borrowed by human science—become our most precise medicines.

Mold may rot bread, but it also rewrote the fate of heart disease.

More than a Best-Seller—A Life Saver

Lipitor:

- Prevented millions of heart attacks and strokes
- Extended lives
- Became affordable worldwide after generics launched
- Cemented statins as the cornerstone of cardiovascular prevention

And it proved something essential:

That great medicines don't always begin with fanfare.

Sometimes, they begin in a forgotten flask, a strange culture, and the stubborn belief that better is possible.

A Legacy Forged in Curiosity

The story of Lipitor is not just about sales or science.

It's about the power of curiosity.

About scientists who look at something small—like mold—and see possibility.

About the patients who lived to see birthdays, weddings, retirements—because someone refused to accept "good enough."

It's about how nature, when paired with human persistence, can save lives at scale.

A Widow's Story: "Why my Husband Lived 20 Years Longer"

One Family's Journey through Medicine, Memory, and the Gift of more Time

He was supposed to die in 1996.

That's what the doctors gently prepared her for.

John, 56, had just suffered his second heart attack. His arteries were severely blocked, his blood pressure unstable, and his LDL cholesterol—"the bad kind," they said—was dangerously high.

His wife, Margaret, sat at his bedside, gripping his hand, trying to memorize the rise and fall of his chest. Trying to imagine a life without his voice at the breakfast table, without his humming in the garden, without him beside her in bed.

But John didn't die in 1996.

Instead, he lived to see his children get married.

He held all four of his grandchildren in his arms.

He took daily walks, planted tomatoes every spring, and renewed their wedding vows on their 50th anniversary.

He lived another twenty years.

And Margaret would tell anyone who asked why:

" Because medicine didn't give up on him. And neither did he."

The Turning Point

In the cardiac ward, after his second heart attack, the doctors revised his medications. They told Margaret and John about a new class of drugs—statins—that aggressively lowered cholesterol and could reduce the risk of another heart attack or stroke by nearly half.

John was skeptical. "Another pill?" he'd said.

But Margaret pressed his hand and said, "Let's try."

He started atorvastatin (Lipitor).

He also took a beta blocker, an ACE inhibitor, and a baby aspirin.

He changed his diet. Walked every morning. Checked in with his doctor every few months.

Within a year, his LDL cholesterol had dropped by over 50%. His blood pressure was under control. He felt stronger.

And most importantly, he didn't land back in the hospital.

The Gift of "More"

Margaret described those years as borrowed time—but not borrowed joylessly.

" We didn't waste a second. We traveled. We laughed more. He learned how to email. We adopted a stray cat. He made pancakes for the grandkids every Saturday."

Every little moment felt like a miracle—because it wasn't supposed to happen.

Without statins, she believes, he wouldn't have made it past 60.

With them—and with early intervention, lifestyle change, and routine care—he lived to 76.

The Quiet Heroism of Prevention

John never felt like a hero.

But every morning when he took that little white pill, he was choosing to stay—for Margaret, for his children, for himself.

" It wasn't dramatic," Margaret said. "It wasn't a transplant or a miracle surgery. It was simple. Small pills, every day. A slow walk. A decision to live."

And that's what made it profound.

Because while some lives are saved in emergency rooms, others are saved around the breakfast table, one pill at a time.

When he Finally went

John passed peacefully, holding Margaret's hand, just weeks shy of their 55th wedding anniversary. He had lived long enough to see the world change, his family grow, and science fulfill its promise.

And at the funeral, as his grandchildren read poems and his daughter played his favorite piano piece, Margaret thought:

"They were all here because he was here.

And he was here because medicine gave us twenty more years."

The Legacy of Longer Lives

John's story is one among millions.

Statins, early intervention, and compassionate care have redefined aging—turning heart disease from a fatal inevitability into a manageable condition.

Behind every prescription is not just a patient—but a family, a community, a future that gets to happen.

And behind every widow like Margaret is a love story that was allowed to last just a little longer.

Chapter 11

ACE Inhibitors: Unlocking the Secret to Blood Pressure

How Snake Venom, South American Legends, and Scientific Tenacity Helped Tame a Silent Killer

High blood pressure was once like a ticking time bomb—one that couldn't be heard, only felt too late.

By the time symptoms appeared—headaches, blurred vision, chest pain—the damage was often done. Arteries thickened. The heart strained. Stroke, heart failure, kidney failure—all crept in quietly, in the shadows.

Doctors had few tools. Some early medications caused fatigue, dizziness, or troubling side effects. Patients hated them. Many stopped taking them. And so hypertension remained what it was always feared to be:

The silent killer.

Then, in the 1980s, came a breakthrough born from venom, blood chemistry, and bold science.

They were called ACE inhibitors.

And they didn't just lower blood pressure—they changed the course of cardiovascular disease forever.

The Clue Hidden in a Snakebite

Long before ACE inhibitors appeared on pharmacy shelves, the answer was hiding in an unlikely place: the venom of the Brazilian pit viper (Bothrops jararaca).

For centuries, Indigenous tribes in South America observed that people bitten by this snake suffered a sudden and dramatic drop in blood pressure—sometimes fatally so.

In the 1960s, Brazilian scientist Dr. Sérgio Ferreira isolated a peptide in the venom that prevented the breakdown of bradykinin, a substance that dilates blood vessels. He sent his findings to Dr. John Vane, a British pharmacologist later known for explaining how aspirin works.

Vane and his team realized this venom peptide blocked an enzyme:

Angiotensin-Converting Enzyme (ACE)—the switch that activated angiotensin II, a hormone that constricts blood vessels and raises blood pressure.

If you could block that enzyme…

You could unwind the body's pressure valve.

The First ACE Inhibitor: From Venom to Victory

Inspired by this, researchers at Squibb Pharmaceuticals (later part of Bristol-Myers Squibb) synthesized a molecule called captopril—approved by the FDA in 1981.

It was the first oral ACE inhibitor.

And it worked.

Captopril lowered blood pressure reliably and safely. But more importantly, it did so by interrupting a key step in the body's internal pressure system—not just calming symptoms, but treating the root cause.

Soon, a wave of ACE inhibitors followed:

- Enalapril
- Lisinopril
- Ramipril
- Benazepril

Each refinement brought better tolerance, longer action, and wider use.

Why ACE Inhibitors were a Game-Changer

ACE inhibitors didn't just reduce blood pressure—they:

- Reduced heart attack and stroke risk
- Slowed the progression of kidney disease, especially in diabetics
- Improved survival after heart failure
- Helped prevent cardiac remodeling, preserving heart function over time

They became the cornerstone of modern cardiovascular therapy.

And unlike earlier medications, they were well-tolerated, widely available, and affordable.

A Patient's Story: From Crisis to Control

Thomas, a 49-year-old bus driver in Detroit, had always brushed off his high blood pressure. "I'm just built that way," he'd joke.

But after a mild stroke, his doctor put him on lisinopril.

Three months later:

- His pressure was down.
- His dizziness was gone.
- He could breathe more easily when climbing stairs.

"I didn't know how bad I felt—until I didn't feel bad anymore," he said.

For the first time, he wasn't afraid of his own body.

Millions Saved, Silently

Since their approval, ACE inhibitors have:

- Been prescribed to hundreds of millions worldwide
- Prevented countless strokes, heart attacks, and kidney failures
- Extended lives—not dramatically, but quietly, consistently, powerfully

They are now recommended in virtually all hypertension guidelines worldwide.

From Snake Venom to Lifesaver

What began as a lethal bite became a chemical blueprint for hope.

What was once feared became a friend.

And what once killed quietly is now being defeated quietly, every day, in millions of lives around the world.

The Legacy of ACE Inhibitors

They didn't make headlines like miracle cures.

They didn't launch cultural movements like antidepressants.

But they changed the arc of public health—giving people more birthdays, more healthy years, more time.

In every grandmother who didn't have a stroke,
In every father who lived to see retirement,
In every patient who walks into a pharmacy for a simple refill—
The legacy of ACE inhibitors lives on.

Snake Venom to Miracle Pills

How a Lethal Bite Inspired One of the Most Life-Saving Classes of Drugs in Medicine

In the jungles of Brazil, a slender, silent predator slithered through the undergrowth—the jararaca pit viper, feared for its bite and deadly precision.

To the people of the Amazon, a bite from this snake was often fatal, but curious in one specific way:

Victims didn't just suffer pain—they experienced a sudden, severe drop in blood pressure, so profound that some would collapse within minutes.

For centuries, this venom was feared.

But science would do what it so often does:

It would turn fear into discovery.

It would take something dangerous—and transform it into hope.

That transformation would lead to ACE inhibitors—a class of miracle pills now taken by millions around the world to prevent heart attacks, strokes, and kidney failure.

A Curious Observation becomes a Clue

In the 1960s, Brazilian pharmacologist Dr. Sérgio Ferreira began studying why the snake's venom caused such dramatic hypotension. What he discovered was astonishing: the venom contained peptides that inhibited an enzyme deep within the body's blood pressure system.

That enzyme?

Angiotensin-Converting Enzyme (ACE)—a critical step in producing angiotensin II, a hormone that constricts blood vessels and raises blood pressure.

The venom blocked this process.

It relaxed the vessels.

Dropped the pressure.

Shut off the body's internal thermostat.

Dr. Ferreira published his findings and sent his work to Dr. John Vane, a prominent British pharmacologist (and future Nobel laureate). Vane recognized what this could mean:

If we could mimic this effect—safely and precisely—we might have a new way to treat high blood pressure at its source.

From Venom to Vitality: The Birth of a New Drug

Inspired by the venom's biochemistry, researchers at Squibb Pharmaceuticals (now part of Bristol-Myers Squibb) synthesized a compound:

Captopril, the world's first ACE inhibitor, approved in 1981.

It was a pharmaceutical first:

- A drug that didn't just mask high blood pressure—but interrupted the internal chemistry behind it.
- A pill that mimicked snake venom—without the bite.

Captopril worked.

Blood pressure dropped.

Patients felt better.

Stroke and heart attack risks declined.

Soon after came enalapril, lisinopril, and others—refinements with longer half-lives and fewer side effects. ACE inhibitors became a global standard of care.

The Lifesaving Impact of the "Venom-Inspired" Pill

ACE inhibitors are now used to treat:

- Hypertension
- Congestive heart failure
- Diabetic kidney disease
- Post-heart attack recovery

They:

- Prevent strokes and heart attacks
- Slow kidney damage in diabetics
- Improve survival in heart failure patients

In total, these medications have saved tens of millions of lives.

They are taken daily by people who will never know they have venom to thank for it.

The Legacy: When Nature Teaches us to Heal

The story of ACE inhibitors reminds us that medicine's greatest innovations often come not from human invention alone—but from listening to nature:

- Penicillin from mold.
- Streptomycin from soil.
- And now—lifesaving heart medication from the venom of a deadly snake.

It's poetic.

What once stopped hearts now helps them beat stronger, longer.

From Poison to Pill: A Modern Medical Miracle

In every small white tablet taken before breakfast…

In every stroke that never happened…

In every heartbeat that lasts longer than it would have…

There lives the unlikely legacy of a creature once feared.

From snakebite to breakthrough, from jungle threat to pharmacy shelf—

Medicine found a miracle in the venom.

Changing the Fate of Generations with Hypertension

How One Breakthrough Turned a Family Curse into a Manageable Condition—and gave the Future a Fighting Chance

For decades, hypertension—high blood pressure—was a shadow that hung over families like a curse.

It didn't strike loudly. It didn't strike early.

But it struck surely, and often fatally.

A father died at 52 from a heart attack.

A grandmother who suffered a massive stroke in her sleep.

A mother on dialysis after kidney failure.

Generations worn down not by war or poverty—but by a disease that crept silently through the arteries, doing its work long before anyone knew it had arrived.

It was a fate passed down like eye color or blood type.

Until one day, science changed the rules.

The Family that Defied its History

The Thompsons were that kind of family.

- Grandpa Joe had a stroke at 60.
- Aunt Louise died in her early 50s after a massive heart attack.
- Cousins whispered about "the pressure problem" running in the bloodline.

So when Eric Thompson, age 38, was told during a routine checkup that his blood pressure was 160/100, he wasn't surprised. Just resigned.

"Figured I'd just go the same way as the rest of them," he said.

But his doctor didn't let it go.

Instead, she pulled out a prescription for lisinopril, an ACE inhibitor, and said,

"That may have been your family's story. But it doesn't have to be yours."

One Pill, One Shift, One Generation Rewritten

Within a few weeks, Eric's pressure dropped. He had more energy. His headaches vanished. His risks plummeted.

Years later, his two children—aware of their family history—got screened early. His daughter, now in her 20s, takes a low dose of medication and works out every morning. His son, inspired by Eric's change, became a nutritionist.

Eric is now 62. He gardens. He hikes. He never had a stroke. Never had a heart attack.

"I broke the cycle," he says. "That was my gift to my kids. They'll bury me old. Not early."

A Public Health Turning Point

What happened in Eric's family is playing out all over the world.

- In the 1970s, fewer than 10% of people with high blood pressure had it under control.
- Today, that number has quadrupled in many countries, thanks to effective and affordable medications—ACE inhibitors, beta blockers, ARBs, and more.

As blood pressure control improved, stroke and heart attack rates dropped, and life expectancy increased.

Hypertension, once a family fate, became a manageable condition.

From Inheritance to Empowerment

Genetics load the gun. But medicine can remove the bullets.

And that's what antihypertensive therapy has done—not just for individuals, but for generations:

- Parents living longer for their children.
- Children growing up without fear.
- Grandchildren born into families where blood pressure is checked—not ignored.

This isn't just prevention. It's reclamation.

Quiet Success, Loud Impact

There's no parade for controlled blood pressure.

No applause for the father who didn't die at 50.

No headlines for the woman who took her meds and didn't have a stroke.

But if you look around, you'll see the impact:

- In weddings attended.
- In birthdays celebrated.
- In generations spared from grief that once seemed inevitable.

The Gift of Interruption

When medicine gave us tools to treat hypertension, it gave families something profound:

The ability to interrupt their own history.

To say: "It stops with me."

To trade resignation for action.

To replace fear with knowledge.

To pass down not disease—but awareness, control, and a longer, fuller life.

Chapter 12

The Birth of IVF and Hormonal Therapies

From Heartbreak to Hope—How Science Helped Create Life where there was None

For countless couples through history, the inability to conceive was more than a medical condition.

It was grief without a funeral, a mourning of children never born, futures never realized, names never whispered aloud.

Infertility was whispered about, often cloaked in shame. For women, it brought silent suffering. For men, quiet guilt. For couples, strain and sorrow. There were no answers. No treatments. Just the aching emptiness of possibility denied.

But in the late 20th century, science gave the world something extraordinary:

A way to create life outside the body.

The birth of in vitro fertilization (IVF) and modern hormonal therapies didn't just open wombs—it opened a new era of reproductive hope.

The First Miracle: A Baby Named Louise

It began, appropriately, with a birth.

In 1978, in Oldham, England, a baby girl named Louise Joy Brown came into the world through what was then an unimaginable technique:

Her mother's egg had been fertilized outside the body—in a glass dish—and then implanted back into her uterus.

She was the world's first "test tube baby."

Behind that birth was a pair of scientific pioneers:

- Dr. Robert Edwards, a physiologist who studied human fertilization

- Dr. Patrick Steptoe, a gynecologist who mastered the technique of egg retrieval using laparoscopy

They battled skepticism, ridicule, and intense ethical controversy. But they persisted.

"We just wanted to help women who wanted children," Dr. Edwards later said.

The Role of Hormonal Therapies: Preparing the Body for Life

IVF alone wasn't enough. For it to succeed, science had to understand—and regulate—the delicate hormonal choreography of ovulation, implantation, and pregnancy.

Enter hormonal therapies, which became the essential backbone of assisted reproduction:

- Clomiphene citrate (Clomid) stimulated ovulation in women with irregular cycles
- Human chorionic gonadotropin (hCG) triggered the final maturation and release of the egg
- FSH and LH analogues controlled ovarian stimulation during IVF cycles
- Progesterone support helped the uterine lining hold onto early pregnancy

These weren't miracle cures. They were carefully timed nudges that coaxed the body into cooperation.

The Revolution that Followed

By the 1980s and 1990s, IVF had transitioned from an experimental to an accessible treatment.

Hormonal therapies improved in precision.

Embryo freezing became viable.

Egg and sperm donation expanded options.

Success rates climbed.

What had once been the privilege of a few became a global movement—and a new branch of medicine: reproductive endocrinology.

The Human Stories: From Waiting to Holding

A couple in India, childless after 12 years of trying, welcomed twins through IVF in 1991. The mother later said, "For the first time in our marriage, the silence was broken—by crying."

A woman in Italy, unable to conceive due to hormonal imbalance, used Clomid and conceived naturally after three months. She wrote in her journal: "I peed on a stick and burst into tears—not from disappointment, but disbelief."

A cancer survivor, who froze embryos before chemotherapy, used hormonal therapy to prepare for implantation. She gave birth to a baby boy she thought she'd never meet.

These stories didn't just represent children.

They represented healing, hope, and wholeness.

A Global Legacy of Life

As of today:

- Over 10 million babies have been born through IVF
- Millions more have been conceived with the help of ovulation induction, hormonal regulation, and egg preservation
- Treatments are now available in over 80 countries, bringing joy to couples, single parents, and LGBTQ+ families alike

Beyond Fertility: Hormonal Therapy's Wider Reach

Hormonal therapies also transformed care for:

- Polycystic ovary syndrome (PCOS)
- Endometriosis
- Early menopause
- Hormone replacement in transgender care
- Prevention of miscarriage in high-risk pregnancies

These medicines helped not only create life but also restore the quality of it for those whose hormones once stood in the way.

The Science of Creation, the Heart of Humanity

It's easy to talk about follicles, FSH levels, and fertilization.

Harder to describe the joy of someone who once thought they'd never hear a child say, "Mommy."

The birth of IVF and hormonal therapies changed science, yes.

But more importantly, they changed fates.

They gave people choice.

They gave love a way to grow.

They gave grief an alternative ending.

Sometimes the greatest breakthroughs aren't about saving life, but making it possible at all.

Unlocking Fertility: The Discovery of Clomiphene, Gonadotropins, and IVF Support Drugs

How Three Classes of Medications Turned the Impossible into Parenthood

Infertility isn't just a medical diagnosis—it's an emotional cliff. Couples sit in silence in waiting rooms, hoping for answers, fearing disappointment. Behind every hopeful calendar marked with ovulation tests and clinic visits is a quiet question:

"Will we ever hold our child?"

For much of the 20th century, the answer was: "We don't know."

Then, three classes of drugs emerged—clomiphene, gonadotropins, and IVF support hormones—that didn't just treat infertility. They created pathways to life.

Together, they changed not just medicine, but millions of family stories.

Clomiphene: The First Fertility Pill

The journey began in the 1950s at Merrell Company (later part of Sanofi), where scientists were developing synthetic estrogens. One compound, clomiphene citrate, behaved unexpectedly: in animals, it stimulated ovulation rather than suppressing it.

Clomiphene worked by tricking the brain. It blocked estrogen receptors in the hypothalamus, leading the brain to believe estrogen levels were low. In response, the body released more FSH and LH, hormones that drive ovulation.

It was subtle. It was elegant.

And it was revolutionary.

In 1967, clomiphene was approved by the FDA under the brand name Clomid. It was the first oral fertility medication. For women with irregular cycles, anovulation, or polycystic ovary syndrome (PCOS), Clomid offered a chance at conception without injections or invasive procedures.

" It gave me control over something that felt hopeless," said one of the first Clomid patients. "And a year later, I had Emma."

Gonadotropins: Precision and Power

For women who didn't respond to clomiphene—or needed more targeted stimulation—another breakthrough arrived: exogenous gonadotropins.

These injectable hormones, derived initially from the urine of postmenopausal women (and later produced via recombinant technology), mimicked the body's own FSH (follicle-stimulating hormone) and LH (luteinizing hormone).

They allowed fertility specialists to:

- Precisely stimulate multiple follicles
- Control timing of ovulation
- Monitor and adjust cycles in real time

Gonadotropins have become essential in IVF protocols, enabling multiple eggs to mature and thereby improving success rates.

Key Developments:

- Human menopausal gonadotropin (hMG) — first derived in the 1950s
- Purified FSH and LH — introduced in the 1980s
- Recombinant FSH (e.g., Gonal-F, Follistim) — launched in the 1990s, offering purity and reliability

These drugs transformed IVF from a long shot to a reproducible, clinical strategy.

hCG: Triggering the Final Step

Once the follicles were mature, the body needed a cue to trigger egg release. That cue came in the form of human chorionic gonadotropin (hCG).

Similar to LH, hCG was used to trigger ovulation—timed precisely before egg retrieval or intercourse. It became the keystone between stimulation and conception.

Progesterone: Helping Life Take Hold

After fertilization—either naturally or through IVF—one last challenge remained: ensuring the embryo implanted and stayed.

Enter progesterone supplementation, often in the form of:

- Vaginal suppositories or gels
- Intramuscular injections
- Oral tablets (less commonly)

Progesterone thickens and stabilizes the uterine lining, supporting early pregnancy until the placenta takes over hormone production.

Without it, many early pregnancies would never take root.

The Synergy that Changed Lives

Together, these therapies created a complete hormonal symphony:

1. Clomiphene or gonadotropins to stimulate ovulation
2. hCG to trigger egg release
3. Progesterone to support implantation

Each step was guided by science—but driven by the deepest human desire: to nurture life.

A New Era of Possibility

By the 1990s and 2000s, these medications had become the foundation of fertility medicine worldwide. They enabled:

- Natural conception in previously untreatable cases
- IVF success rates to climb from <10% to >40% per cycle in some populations
- Fertility preservation for cancer patients and older parents
- Inclusive family-building for LGBTQ+ couples and solo parents

The Numbers Tell the Story

Since their introduction:

- Over 10 million babies have been born with the help of fertility drugs
- An estimated 1 in 6 couples now seek fertility support globally
- These therapies are part of millions of success stories—quiet victories that unfold in bedrooms, nurseries, and playgrounds

Behind Every Syringe, a Story

Behind every injection is:

- A woman who gave herself daily shots with trembling hands
- A partner who whispered, "We'll try again" after a failed cycle
- A grandmother cradling a baby she thought she'd never meet

These medicines did not just change fertility—they restored hope.

A Legacy of Creation

The discovery of clomiphene, gonadotropins, and IVF support drugs turned heartbreak into healthcare and longing into life.

They didn't just offer a chance to conceive.

They offered a chance to believe again.

" When science meets hope," one IVF nurse said, "that's when miracles are born."

From Infertility to Joy: "Meet the Child we couldn't have"

One Couple's Journey Through Despair, Determination—and the Day They Finally Held Hope in Their Arms

For Anita and Raj, the journey to parenthood wasn't marked by baby showers or nursery shopping.

It was marked by doctor visits, hormone shots, silent prayers, and empty arms.

Married for eight years, they had built a life full of love—but not laughter. Not the kind that came from little feet and bedtime stories. Every month was another heartbreak. Every family function another reminder. Every "When are you having children?" was a question they'd already asked themselves a hundred times that week.

"It wasn't just about wanting a baby," Anita later said. "It was about wondering if we were somehow incomplete."

Their medical diagnosis: unexplained infertility.

Their emotional diagnosis: grief with no clear end.

The First Step: Science Offers a Hand

Their doctor explained the options:

They began with clomiphene citrate (Clomid) to stimulate ovulation.

Nothing.

Next came gonadotropin injections, carefully monitored by ultrasound Hormones surged. Emotions did too.

They tried intrauterine insemination (IUI).

Then IVF. Then another round.

Then a break.

"We were exhausted—not just financially, but spiritually," Raj recalled. "We were living in two-week intervals. Hope. Wait. Disappointment. Repeat."

They had one frozen embryo left.

It would be their last try.

The Call that Changed Everything

Anita remembers where she was—on the couch, not expecting anything—when the clinic called.

"You're pregnant."

She didn't scream. She didn't cry. She whispered, "Really?"

Because after years of trying, of dreaming, it didn't feel real yet.

Every day of that pregnancy felt like a miracle wrapped in caution. Every ultrasound, every heartbeat, every little flutter inside her reminded them: this is happening.

The Child they couldn't have

Nine months later, their daughter Maya was born.

"She didn't cry right away," Anita said, "and I panicked—until she opened her mouth and let out the most beautiful sound I've ever heard."

They held her.

They looked at each other.

They cried—not because they were sad, but because their arms weren't empty anymore.

More than a Baby—A Victory

Maya wasn't just a child. She was:

- A vial of hormone injections
- A stack of medical bills
- A whispered promise not to give up
- A lifetime of longing turned into laughter

"She's the child we couldn't have," Raj said, cradling her. "But here she is—breathing, smiling, changing everything."

Science made her Possible. Love made her Inevitable

Behind Maya's birth were:

- The researchers who discovered clomiphene
- The scientists who developed injectable gonadotropins
- The embryologists who watched a cluster of cells divide
- The doctors and nurses who held Anita's hand
- The mother and father who refused to stop believing

She was built not just by biology, but by hope, perseverance, and the science of miracles.

Millions of Miracles just Like her

Today, Maya is one of over 10 million children worldwide born through assisted reproductive technologies. Behind each one is a story like Anita and Raj's—quiet, painful, brave, and triumphant.

They are the "children we couldn't have"—until science said otherwise.

The Legacy of One Little Girl

Maya will never know the depth of what her birth meant—not until she has children of her own.

But every time she laughs, takes a step, says "I love you,"

She proves something her parents already know:

"Some miracles take longer to arrive. But they're worth every moment of waiting."

PART IV: PHARMA IN THE FACE OF CRISIS (1980s - 2000s)

Chapter 13: AZT and the War on HIV/AIDS

Chapter 14: Triple Therapy and the Turning Point in HIV

Chapter 15: Cancer Breakthroughs: From Chemotherapy to Targeted Therapy

Chapter 16: Vaccines: Saving Millions Quietly

Part IV

Pharma in the Face of Crisis (1980s - 2000s)

How an Industry Stepped into the Fire—and began to Rebuild Trust, One Breakthrough at a Time

The late 20th century wasn't just a time of discovery for pharma—it was a time of reckoning.

It was a time of pandemics, pain, and public scrutiny.

Of headlines filled with tragedy.

Of pressure not just to produce, but to act, to respond, to deliver hope under siege.

From the devastation of the HIV/AIDS epidemic, to global vaccine shortages, to rising waves of chronic illness and inequality, the pharmaceutical industry stood at a crossroads.

Would it be remembered for patents—or for people?

This is the story of how pharma, under fire and facing enormous expectation, delivered some of its most profound breakthroughs—and began to reshape its relationship with the world.

The HIV/AIDS Epidemic: Pharma Meets a Pandemic

When HIV began its spread in the early 1980s, it didn't come with clear symptoms, known treatments, or even a name.

It brought death.

It brought stigma.

It brought silence.

Pharma was slow to act. Public funding was even slower. And by the time attention caught up to urgency, tens of thousands had died.

But then something extraordinary happened. Under pressure from activists, patients, scientists, and society, pharma pivoted into overdrive.

- In 1987, AZT (zidovudine) became the first antiretroviral drug approved for HIV. It wasn't perfect—but it was a start.
- By the 1990s, companies like Glaxo, Merck, Roche, and Bristol-Myers Squibb launched a new generation of drugs targeting protease and reverse transcriptase enzymes.
- In 1996, HAART (Highly Active Antiretroviral Therapy) became the standard of care, transforming HIV from a death sentence into a chronic, manageable condition.

Patients who had been preparing to die instead began planning for the future.

"My meds gave me years I didn't expect. They gave me the chance to live long enough to fall in love again," one survivor later said.

The Global Health Wake-Up Call

But while the West began to breathe easier, the crisis deepened in Africa and the Global South, where millions couldn't afford lifesaving drugs.

This sparked one of the most important ethical pivots in pharma history:

- Generic licensing agreements and tiered pricing became tools for access.
- Partnerships with WHO, UNAIDS, and nonprofits brought ARVs to millions.
- Companies like Cipla, Gilead, and Abbott began crafting global health strategies, not just marketing plans.

Pharma learned something enduring:

Breakthroughs don't matter if they don't reach the people who need them most.

The Painkiller Paradox

While pharma was fighting one global epidemic, it was inadvertently laying the groundwork for another: the opioid crisis.

In the 1990s, pain was rebranded as the "fifth vital sign." Patients demanded relief. Physicians were urged to prescribe more compassionately.

Enter powerful opioid painkillers—morphine derivatives, oxycodone, and hydrocodone. Some were marketed as "non-addictive when used for pain."

They weren't.

By the early 2000s, overprescription, addiction, and diversion had taken root. The same industry that brought relief now faced blame. Lives were saved—but others were destroyed.

The opioid crisis would become pharma's deepest moral wound, prompting lawsuits, regulation, and a reckoning that still unfolds today.

Vaccines, Pandemics, and Preparedness

As the world entered the 2000s, new threats emerged:

- Hepatitis C and B
- Tuberculosis resurgence
- Influenza pandemics
- SARS in 2003

Pharma met them with renewed urgency:

- Hepatitis B vaccines became standard in infants.
- Hepatitis C antivirals were pioneered—paving the way for later cures.
- The Tamiflu (oseltamivir) stockpiling effort began in earnest.
- Pneumococcal and rotavirus vaccines spread globally, saving millions of children's lives.

This was the beginning of pandemic preparedness—not perfect, but a foundation that would prove critical in the decades to come.

Chronic Illness and the Rise of Maintenance Medicine

As infectious threats were being addressed, a new crisis crept in quietly: the global rise of non-communicable diseases—heart disease, diabetes, cancer, and depression.

Pharma responded with a wave of maintenance medications:

- Statins for cholesterol
- ACE inhibitors and beta blockers for hypertension
- Metformin and insulin analogs for diabetes
- SSRIs and SNRIs for depression

These weren't flashy cures. But they turned millions of life-shortening diagnoses into manageable conditions. They allowed people to live longer, better, more functional lives.

The success was quiet—but revolutionary.

Trust Eroded, Then Rebuilt

The late 1990s and early 2000s also saw growing distrust in pharma:

- Pricing scandals
- Marketing ethics
- Hidden side effects
- Corporate greed headlines

But it also saw the beginnings of change:

- Transparency laws for clinical trials
- Patient advocacy partnerships
- Global access initiatives
- Investment in rare disease research and neglected tropical diseases

Pharma began to realize: its role wasn't just to produce drugs, but to act responsibly, listen compassionately, and serve humanity as a whole.

A Period of Crisis—and Conscience

The 1980s to 2000s was pharma's trial by fire.

It saw both its brightest moments—like antiretroviral therapies and lifesaving vaccines—

and its darkest missteps—like the seeds of the opioid epidemic and access inequality.

But it also proved something essential:

In crisis, pharma is capable of rising—not just to the scientific challenge, but to the human one.

It learned that innovation is not enough unless it is ethical, equitable, and accessible.

Chapter 13

AZT and the War on HIV/AIDs

The First Strike against a Merciless Virus—and the beginning of Hope for Millions

In the early 1980s, hospitals began to see something strange.

Young men, previously healthy, were showing up with rare infections and aggressive cancers. Their immune systems had collapsed. Many died within months. No one knew why.

They called it GRID at first—Gay-Related Immune Deficiency.

Then, as women and children began to fall ill, the truth became undeniable:

This wasn't a lifestyle disease.

It was a global viral epidemic.

The name changed to AIDS. The virus was called HIV.

And the prognosis was bleak.

In the face of widespread stigma, government indifference, and rising death tolls, the world desperately needed a miracle.

In 1987, that miracle came in the form of a small white pill called AZT—the first drug ever approved to fight HIV.

It wasn't perfect. It wasn't even gentle.

But it was the first sign that HIV could be fought.

And that people with AIDS could live, not just die waiting.

The Virus that Terrified the World

HIV (Human Immunodeficiency Virus) attacks CD4 T-cells, the very immune cells that defend the body. Left untreated, the virus replicates until the immune system collapses, opening the door to deadly opportunistic infections.

By the mid-1980s, the epidemic had:

- Killed tens of thousands in the U.S. alone
- Spread across Africa, Asia, and Europe
- Carried a nearly 100% fatality rate

There were no treatments. Only isolation, experimental trials, and heartbreak.

Funeral homes refused bodies. Families disowned children. Patients were abandoned—by friends, by society, even by medicine.

And yet, in the middle of this storm, activists, scientists, and patients rose up.

An Old Drug, a New Hope

AZT (azidothymidine) wasn't designed to fight HIV. It had been shelved in the 1960s as a failed cancer drug.

But in 1984, researchers at Burroughs Wellcome (now part of GSK) tested it in vitro against the newly identified HIV virus.

The results were astonishing:

AZT stopped HIV from replicating by inserting itself into the virus's DNA and terminating the chain—a form of antiretroviral therapy.

The urgency was so great that AZT went from testing to approval in just 25 months—the fastest drug approval in FDA history at that time.

On March 19, 1987, the FDA approved AZT under the name Retrovir.

The First Line of Defense

AZT gave patients something they hadn't had before: time.

- It slowed disease progression
- Raised CD4 counts temporarily
- Helped patients live longer than expected

For many, it was the first taste of hope.

"I took AZT. I got six more years with my partner. Six years we weren't supposed to have," one survivor recalled.

But the drug wasn't easy.

- It required every 4-hour dosing, day and night
- It caused anemia, nausea, and fatigue
- Over time, it stopped working alone—the virus adapted

Still, for those dying with nothing, AZT was life.

The Activists who Pushed the System

While AZT marked progress, it also sparked outrage over:

- High pricing ($10,000/year—a fortune in the 1980s)
- Limited access
- Concerns over trial ethics and representation

Enter ACT UP, TAG, and other activist coalitions. They demanded:

- Faster approval processes
- Inclusive clinical trials
- Compassionate-use programs
- Access to experimental drugs for dying patients

Their protests didn't just shake pharma and the FDA—they reshaped how drugs are developed and regulated, especially for life-threatening diseases.

AZT was the flashpoint.

But what followed was a revolution in patient advocacy and drug access.

Combination Therapy: AZT's Legacy

By the mid-1990s, researchers realized that AZT alone wasn't enough. HIV mutated too quickly.

But when AZT was combined with other antiretrovirals—especially protease inhibitors and reverse transcriptase inhibitors—the results were transformative.

In 1996, the world entered the age of HAART (Highly Active Antiretroviral Therapy).

Suddenly:

- Viral loads dropped to undetectable levels
- Patients went back to work, to school, to life
- Mother-to-child transmission dropped dramatically

AZT became a foundation stone in a multi-drug regimen that turned HIV into a manageable chronic illness.

The Numbers Tell the Story

Since its approval, AZT has:

- Prolonged the lives of millions globally
- Played a key role in reducing HIV transmission to newborns
- Served as a model for drug repurposing under crisis conditions
- Reminded pharma that urgency, ethics, and access must walk together

From Death Sentence to Chronic Condition

Today, HIV is still incurable.

But it is no longer unbeatable.

People on modern regimens—including descendants of AZT—can expect near-normal life expectancy.

They fall in love. Raise families. Grow old.

AZT was never the final answer.

But it was the first answer.

And that changed everything.

The Pill that Bought Time—and Sparked a Movement

It came from a dusty shelf.

It left side effects that were brutal.

But AZT did something extraordinary:

It gave people time to live long enough for better drugs to be developed.

It turned patients into activists.

Activists into policy changers.

And a terrified, grieving community into a force that would transform global healthcare forever.

The Urgency, Activism, and Discovery of the First Antiretrovirals

How a Community Under Siege Sparked Science into Action and Changed Medicine Forever

By the early 1980s, the streets of New York, San Francisco, and Los Angeles were haunted not by violence or war—but by something far more insidious:

A virus with no name.

A disease with no cure.

And a silence that killed faster than the illness itself.

Young men were dying of pneumocystis pneumonia, Kaposi's sarcoma, and sudden immune collapse. Their bodies were wasting away in hospital beds. Their friends held their hands while wearing gloves and masks. Their partners were barred from funerals. And the medical world had no answers.

This was the beginning of the AIDS crisis.

And it was as much a test of compassion as it was of science.

What followed was a storm of urgency, activism, and a scientific race that gave birth to the first antiretroviral drugs—a breakthrough not just in medicine, but in human rights, research ethics, and the relationship between pharma and the public.

A Pandemic that wasn't Treated Like One

When the first cases appeared in 1981, HIV/AIDS was dismissed by many as a "niche" illness. It affected marginalized communities—gay men, injection drug users, sex workers. Governments turned a blind eye. Funding was minimal. Compassion was scarce.

Patients were dying in silence.

Physicians were powerless.

And pharmaceutical companies had no roadmap, no precedent, and no time.

But while institutions hesitated, people didn't.

The Birth of a Movement: Patients as Pioneers

In hospital wards, bedrooms, and street protests, people with AIDS and their allies organized, educated, and mobilized:

- ACT UP (AIDS Coalition to Unleash Power) staged die-ins and stormed FDA headquarters.
- GMHC (Gay Men's Health Crisis) became a lifeline for thousands.
- Project Inform, Treatment Action Group, and others demanded:
- Faster drug approvals
- Expanded clinical trial access
- Representation and dignity in research

These weren't just protests. They were public health interventions.

" We were dying, and no one was doing enough," one ACT UP member later said. "So we learned the science and demanded the science move."

Patients became experts in immunology, pharmacology, and clinical trial design. And they forced the world to listen.

The Scientific Rush: From Lab Bench to Bedside

Pharmaceutical companies, facing pressure and rising urgency, turned to their shelves. In dusty archives, they found AZT (azidothymidine)—a failed cancer drug from the 1960s.

Researchers at Burroughs Wellcome tested it against HIV in 1984.

It worked. It stopped HIV replication in lab cells.

In record time, clinical trials were launched. By 1986, AZT trials were underway. By March 1987, AZT became the first drug approved for treating AIDS.

It wasn't perfect—expensive, toxic, and only temporarily effective—but it changed the game.

It was the first antiretroviral.

More Drugs Followed—Fast

AZT had shown the virus was chemically vulnerable.

Now pharma companies rushed to develop more potent, more tolerable options:

- Didanosine by Bristol-Myers Squibb
- Zalcitabine by Roche
- Lamivudine by BioChem Pharma
- Stavudine by Yale and licensed to Bristol-Myers

Each built on the model: reverse transcriptase inhibitors that blocked the enzyme HIV used to convert its RNA into DNA.

By the early 1990s, patients were taking combinations of these drugs—doubling and tripling their pill burdens, but also doubling and tripling their chances of survival.

Activism Shapes Regulation

These first drugs didn't arrive just because of science. They arrived because patients demanded a new research paradigm:

- Parallel track access for those too sick for clinical trials
- Accelerated approval pathways for life-threatening diseases
- Inclusion of people with HIV in trial design and review boards

The FDA changed.

Pharma changed.

The ethics of drug development changed—forever.

The Cost of Progress

The early antiretrovirals were not easy.

They were toxic, expensive, and often failed in the long term.

But they bought time.

And for many, that time was everything.

Time for new drugs.

Time to fall in love.

Time to watch a niece graduate.

Time to outlive a death sentence.

A Foundation for the Future

The drugs that followed—protease inhibitors, integrase inhibitors, and HAART—stood on the shoulders of this first wave.

Without AZT, ddI, ddC, and their peers, there would be no Truvada, no Biktarvy, no U=U (Undetectable = Untransmittable).

The first antiretrovirals weren't the cure.

But they were the proof that HIV could be fought—and that patients had the power to shape their own treatment future.

The Lesson that Echoes On

The war on HIV/AIDS taught us that:

- Urgency must be matched by action
- Science must be informed by those it aims to help
- Medicine is not just molecules—it's humanity in crisis, demanding compassion

In the face of fear, activism, and death, the world didn't just find a drug.

It found a voice.

A movement.

A future.

"I Lived Because of AZT": Voices from the Epidemic

The Drug wasn't Perfect, but it bought Time, It brought Hope, and it gave the World Back to People who were Already Saying Goodbye.

In the early years of the HIV/AIDS epidemic, being diagnosed was like being handed a countdown clock. You didn't know how long you had—only that your body, once a fortress, was now a battlefield. And you were losing.

The diagnosis brought fear, shame, and silence. Funeral plans were made. Friends and lovers were buried. Parents outlived their children.

Then, in 1987, a glimmer of hope arrived—a pill called AZT.

It wasn't a miracle cure. It had side effects that knocked some people flat. But for many, it slowed the virus, lifted despair, and most importantly—it gave them time to live.

Here are their stories.

Michael, 28, San Francisco, 1988

"When I started AZT, I weighed 105 pounds. I had thrush. I could barely eat. My partner had died six months earlier. I was ready to go too."

He wasn't supposed to see 1989. But AZT gave his immune system just enough strength to bounce back.

"It didn't make me feel better overnight. But slowly, I started walking again. Eating. Laughing. I got a dog. I started volunteering at the AIDS hospice I thought I'd die in."

Years later, Michael would be among the first to receive combination therapy. But he credits AZT for keeping him alive long enough to get there.

Vanessa, 35, Harlem, 1992

"I was a mother with two kids. I didn't use drugs. I wasn't gay. So when the doctor told me I was HIV positive, I thought he was lying."

The virus didn't discriminate, even if the world did. Vanessa's church turned its back on her. Family whispered.

"When I started AZT, it gave me back control. It said, 'You're not done yet.'"

She lived to see her children graduate college. She now leads a support group for women living with HIV.

James, 42, Los Angeles, 1990

"The thing about AZT wasn't just the drug. It was the message: that we weren't invisible anymore."

James remembers lining up at his clinic with dozens of other patients—all gaunt, all scared. But the day they got their prescriptions, something changed.

"We looked each other in the eye for the first time in months. Because AZT said: someone's working on this. You matter."

He lived long enough to marry his partner in 2015. He keeps his first AZT bottle in a shadow box by their bed.

Rita, 29, Nairobi, 1997

"When my husband died of AIDS, everyone in the village stopped coming to our home. When I tested positive, I thought my children would be orphans."

Through a partnership between her government and an international aid group, Rita received AZT as part of a mother-to-child prevention program.

"My baby was born HIV negative. AZT didn't just save me—it saved her."

She now works as a peer educator, helping other women get tested, treated, and supported.

The Power of "Just Enough Time"

AZT didn't erase the virus. It didn't restore full health. It caused fatigue, anemia, and confusion. It wasn't perfect.

But for many, it gave just enough time to:

- Attend one more Pride march
- Reconcile with a parent
- Start combination therapy
- Witness the rise of U=U
- Fall in love again
- Simply live

"AZT didn't save me forever," one survivor said. "It just saved me long enough. And that made all the difference."

A Pill and a Promise

To those who took AZT, it was more than a drug. It was a sign that:

- The world was listening
- Science was moving
- They hadn't been forgotten

It marked the shift from helplessness to hope.

The Legacy in Every Life Saved

Today, AZT is rarely used on its own. It's been replaced by safer, more effective regimens. But its legacy lives in:

- Every antiretroviral combination
- Every child born HIV-negative
- Every long-term survivor
- Every activist who once said, "We deserve to live"

"I lived because of AZT," said James. "And because I lived, I got to see how far we've come."

Chapter 14

Triple Therapy and the Turning Point in HIV

How Three Drugs, One Idea, and Countless Voices Transformed a Death Sentence into a Chronic Condition

For over a decade, HIV meant goodbye.

Goodbye to partners, to dreams, to futures never lived.

The virus didn't just take lives—it stripped people of time, dignity, and hope.

Single drugs like AZT offered glimpses of survival. But the virus adapted fast. It mutated, dodged treatments, and returned stronger. People developed resistance. What started as a promise too often ended in pain.

By the mid-1990s, it was clear:

We weren't winning. Not yet.

Then came a scientific shift. A bold idea. A new strategy:

Triple therapy.

Three drugs. One coordinated assault on the virus.

It was the turning point in the AIDS epidemic—and it changed everything.

The Science behind the Breakthrough

HIV is a retrovirus, which means it integrates its genetic material into a person's DNA. And it's a master of mutation—copying itself imperfectly, evolving with every replication.

That's why monotherapy (single-drug treatment) couldn't hold it off.

Researchers realized that to stop HIV in its tracks, they needed to:

- Hit multiple stages of the virus's life cycle
- Use drugs with different mechanisms of action
- Make it harder for the virus to escape through mutation

The solution was a combination antiretroviral therapy—specifically, a regimen of two nucleoside reverse transcriptase inhibitors (NRTIs) and a protease inhibitor.

Together, they became known as HAART—Highly Active Antiretroviral Therapy.

1996: The Year Everything Changed

The game-changing moment came at the 11th International AIDS Conference in Vancouver in July 1996.

The buzz in the corridors was electric. For the first time, doctors shared clinical trial data showing that triple therapy could suppress HIV to undetectable levels—not just temporarily, but sustainably.

People who had been bedridden were regaining strength.

Hospitals that had overflowed with AIDS patients began to empty.

Obituaries stopped appearing weekly in community newsletters.

"It was like someone had turned the lights back on," one physician said. "People stopped dying."

The Human Impact: From Despair to Defiance

Mark, diagnosed in 1991, had already lost 30 pounds and said goodbye to his family. He was preparing to die.

Then his doctor offered him a new triple-drug combination: AZT, 3TC, and indinavir.

"Within weeks, my fevers stopped. I could eat again. I stopped planning my funeral. I started planning a vacation."

He lived to see 9/11, the legalization of same-sex marriage, and his 60th birthday.

Protease Inhibitors: The Power behind the Punch

At the heart of triple therapy were protease inhibitors—a class of drugs that blocked an essential viral enzyme needed to produce mature, infectious HIV.

- Saquinavir (the first in 1995)
- Ritonavir
- Indinavir
- Nelfinavir

These drugs stopped HIV assembly, slashing viral loads and giving the immune system room to recover.

When paired with NRTIs like AZT, 3TC, or d4T, they became the backbone of HAART.

The Results were Immediate—and Historic

By the late 1990s:

- AIDS-related deaths in the U.S. dropped by more than 60%
- HIV hospital wards closed
- People with HIV began living near-normal lifespans
- The term "AIDS patient" quietly gave way to "long-term survivor."

It was nothing short of a revolution.

From Death Sentence to Chronic Illness

Triple therapy didn't cure HIV—but it controlled it.

- Viral loads became undetectable.
- CD4 counts rebounded.
- Quality of life improved dramatically.

For many, the goal of treatment shifted from "survive" to "thrive."

"I stopped being afraid of the next test result," said Vanessa, a patient since 1988. "I started making plans."

New Challenges, New Responsibilities

The success of triple therapy brought new questions:

- How do we make treatment affordable in the developing world?
- Can we maintain strict adherence—30+ pills a day, often with food restrictions?
- What about side effects: lipodystrophy, nausea, liver strain?

But over time, formulations improved.

Doses decreased.

Side effects were managed.

And pharma responded with fixed-dose combinations, including single-tablet regimens.

What began as three separate pills became, eventually, one pill, once a day.

The Global Impact

Triple therapy inspired:

- The creation of PEPFAR, the largest global health initiative in history
- The launch of The Global Fund to Fight AIDS, Tuberculosis and Malaria
- Affordable access through generics, NGO partnerships, and tiered pricing

Today, over 28 million people around the world receive antiretroviral therapy.

A Turning Point that Saved Tens of Millions

Triple therapy didn't just rewrite medical textbooks.

It rewrote people's lives.

It gave back weddings, careers, friendships, second chances.

It turned HIV from the virus that stole decades into the reason people got to keep them.

And it all began with one idea:

"If one drug isn't enough—try three."

HAART and the Life Extension Revolution

How One Breakthrough gave Millions not just more Time—But a Future to Live into

In the early 1990s, a diagnosis of HIV was a cruel kind of calendar.

Doctors spoke in months, not years.

Patients counted T-cells like grains of sand slipping through the hourglass.

Life revolved around what could be done before the immune system collapsed.

Then, almost suddenly, everything changed.

In 1996, the arrival of HAART—Highly Active Antiretroviral Therapy—didn't just improve outcomes.

It rewrote the course of the epidemic.

It triggered a life extension revolution.

For the first time, HIV wasn't a death sentence.

It was a manageable chronic condition.

And the goal of treatment shifted—from survival to quality of life and longevity.

What is HAART?

HAART refers to the use of three or more antiretroviral drugs from at least two different classes, usually:

- Two NRTIs (nucleoside reverse transcriptase inhibitors)
- One PI (protease inhibitor) or NNRTI (non-nucleoside reverse transcriptase inhibitor)

This multi-pronged attack:

- Reduces viral load to undetectable levels
- Prevents mutation and drug resistance
- Allows the immune system to recover and function again

It was the first time medicine found a way to keep pace with HIV's relentless adaptability.

The First Signs of a Future

When HAART was introduced, patients already on the brink of AIDS—some on hospice care—saw stunning recoveries.

- Viral loads dropped 100-fold, then 1,000-fold.
- CD4 counts rebounded.
- Opportunistic infections slowed or disappeared.

In some hospitals, AIDS wards were dismantled— not because the epidemic ended, but because people weren't dying anymore.

" It felt like science pulled me back from the grave," said Daryl, diagnosed in 1992 and still alive today.

Extending Life—and Improving it

HAART didn't just add years. It added:

- Productivity: People returned to work, finished school, and started families.
- Stability: Treatment regimens became predictable and integrated into daily life.
- Empowerment: Patients were no longer passive recipients of care, but active participants in long-term wellness.

Longevity wasn't a luxury—it became expected.

The Numbers Tell the Story

Before HAART:

- Life expectancy post-diagnosis: less than 10 years, often far less

After HAART:

- Life expectancy of a person diagnosed in their 20s: up to 70+ years, if treated early and consistently
- Mortality from AIDS-related illnesses: plummeted by over 80% in many countries within a few years of HAART adoption

Today, with consistent treatment, many people with HIV live as long as those without it.

The Psychological Turning Point

This extension of life came with a radical shift in identity.

- From "I'm dying" to "I'm living."
- From secrecy and stigma to visibility and resilience.
- From short-term coping to long-term planning.

People who had once cashed out retirement plans and said goodbye to loved ones were now:

- Applying for mortgages
- Starting businesses
- Advocating for global access
- Raising children

"I had to learn how to live again," said Angela, diagnosed in 1990. "HAART gave me time—but also responsibility. I was no longer dying. I was aging."

The Role of Pharma and Global Health

Pharmaceutical innovation surged in response to HAART's success:

- New drug classes: Integrase inhibitors, entry inhibitors, and long-acting injectables
- Fixed-dose combinations: Single pills with three drugs
- Generics and global licensing: Expanded access across Africa, Asia, and Latin America

Programs like PEPFAR and the Global Fund helped bring HAART to over 28 million people worldwide.

What began as a treatment for the privileged became a lifesaving standard for the world.

A New Kind of Longevity Movement

HAART also inspired broader life-extension thinking:

- Early testing and universal treatment
- Preventive medicine and lifestyle integration
- Focus on comorbidities: cancer, diabetes, cardiovascular health in aging HIV patients
- HIV care became a model for chronic disease management

It showed that targeted, sustained, adaptive therapy could not only preserve life—but optimize it.

Legacy and Lessons

HAART redefined what it meant to live with a virus.

It proved that:

- Death isn't inevitable if science is mobilized.
- Life can be restored if dignity is prioritized.
- The line between illness and wellness is often a pill away.

"I didn't just survive. I outlived fear," said Kevin, 66, who has been on HAART since the beginning. "And I'm still not done."

From Death Sentences to Chronic Conditions

How Breakthrough Medicines Turned the World's most Feared Diagnoses into Manageable Lifelong Journeys

For much of history, a diagnosis of diseases like HIV, cancer, diabetes, multiple sclerosis, or heart disease meant one thing:

Get your affairs in order.

These were death sentences—words that collapsed futures into funerals.

There was no roadmap, no second act. Only mourning that began even before the body gave out.

But medicine, like life, is resilient.

And across the last five decades, something remarkable happened:

Science didn't just delay death—it redefined disease itself.

Many of the world's most terrifying conditions were transformed into chronic, manageable illnesses—the kind people now live with, not die from.

The Turning Point: When Medicine Stopped Fighting Battles and Started Winning Wars

This revolution wasn't loud. It didn't always happen in one moment.

It was built molecule by molecule, trial by trial, patient by patient.

And it happened across many fronts:

HIV/AIDS

Once a near-certain death sentence in the 1980s, HIV is now a chronic illness with normal life expectancy—thanks to triple therapy, HAART, and integrase inhibitors. Patients live full lives, love, marry, age. And with undetectable viral loads, they do not transmit the virus.

Cancer

Not always a death knell anymore.

Targeted therapies, immunotherapies, and maintenance drugs have turned many cancers—like chronic myeloid leukemia (CML) and breast cancer (ER+, HER2+)—into long-term conditions.

Patients are monitored, adjusted, and continue living.

Diabetes

Insulin once meant a short, difficult life. Now, with modern analogs, GLP-1 receptor agonists, and digital glucose monitoring, people with diabetes can live as long and as fully as those without it—with the right care.

Cardiovascular Disease

Beta blockers, ACE inhibitors, statins, and anticoagulants have made heart failure, hypertension, and high cholesterol chronic, not catastrophic.

Millions now survive heart attacks and strokes and go on to live productive, healthy lives.

Multiple Sclerosis and Rheumatoid Arthritis

Once disabling and progressive, these autoimmune diseases now have biologic therapies that reduce relapses, slow progression, and preserve quality of life.

The Definition of "Living With" Changed Forever

Before this transformation, being a patient meant waiting to die.

Now, it means planning to live.

People go back to work.

They fall in love.

They raise children.

They travel.

They write books, run businesses, mentor others.

" I take meds every day," one HIV-positive father said, "but what I really manage is my future."

The New Medical Contract: Partnership Over Pity

This shift also redefined the patient-physician relationship:

- Doctors now speak in terms of goals, not limitations.

- Treatment is about control, not cure—and that's enough.
- Patients are empowered to track, choose, and collaborate.

The body is no longer a battlefield—it's a habitat worth tending.

Quality of Life became the New Metric

Survival was once the only measure of success. Now, we ask:

- Can the patient work?
- Are they in pain?
- Do they feel whole?
- Can they dream, plan, laugh?

Medicine now focuses not just on extending life, but on improving it—day by day.

The Role of Pharma: Quietly Powering Possibility

Behind this revolution are decades of pharmaceutical breakthroughs:

- Drugs that control disease at the molecular level
- Therapies that adapt over time
- Sustained delivery systems and long-acting injectables
- Combinations that reduce pill burden and side effects
- Generics and global partnerships that make access possible

What began in sterile labs became the heartbeat of millions of real lives.

From Helplessness to Hope

Once, people whispered their diagnoses.

Now, they post their pill schedules, share fitness goals, join support groups, and write memoirs

They survive—and they thrive.

" I have cancer," one woman said. "But it doesn't have me."

The Future is Chronic—and that's a Good Thing

As life expectancy increases and medicine advances, more of us will live with chronic conditions.

And that's no longer a tragedy. It's a triumph.

It means:

- More birthdays

- More first steps
- More books written and read
- More decades given to people who nearly lost them

A New Era of Health, Defined by Persistence

From death sentences to chronic conditions—this is not just a shift in medicine.

It's a shift in what we believe is possible.

We no longer ask, "How long do I have?"

We ask, "How well can I live?"

And that question—answered daily by millions—is the legacy of modern medicine's greatest promise:

To give life back. And not just for a moment, but for good.

Chapter 15

Cancer Breakthroughs: From Chemotherapy to Targeted Therapy

How Medicine Moved from Blunt Force to Precision Tools—and gave Cancer Patients Back their Futures

Cancer used to be whispered, not spoken.

A shadow diagnosis. A closed-door conversation. A word avoided because it so often ended with another: "Terminal."

The treatments were brutal. The outcomes are uncertain. And for most of the 20th century, cancer care meant a difficult choice: endure a body-breaking assault—or prepare to say goodbye.

But across decades of relentless research and countless clinical trials, medicine changed course. It evolved.

We moved from scorched-earth chemotherapy to pinpoint precision.

From "kill everything" to "target the enemy and spare the rest."

This is the story of that transformation. A story of pain, progress, and hope rebuilt cell by cell.

The Era of Chemotherapy: The First Line of Offense

The roots of chemotherapy were grim: World War II.

Doctors observed that soldiers exposed to mustard gas had suppressed bone marrow and white blood cells.

What was once a chemical weapon became a clue:

Could this effect be used to kill rapidly dividing cells—like cancer?

In the 1940s and 1950s, early chemotherapies like nitrogen mustards, methotrexate, and 5-FU were born.

They weren't precise. They attacked all fast-growing cells—including those in the gut, hair follicles, and bone marrow.

But for the first time in medical history, some cancers—especially childhood leukemias and lymphomas—could be pushed into remission.

It wasn't a cure. But it was a chance.

And that mattered more than anything.

The Harsh Realities of the "Blunt Force" Era

Chemotherapy saved lives. But it also:

- Caused severe side effects: nausea, hair loss, infections, organ damage
- Was physically and emotionally grueling
- Often brought temporary remissions, followed by relapse

Still, patients persisted.

And researchers asked a critical question:

Can we be more accurate?

The Birth of Targeted Therapy: Finding the Lock, Making the Key

As the genetic revolution unfolded in the 1970s and 1990s, scientists learned that not all cancers are the same.

Even within one type—say, breast cancer or lung cancer—there were subtypes, each driven by specific genetic mutations or protein expressions.

If cancer was a lock, these mutations were keys.

Could we make drugs that only fit those locks?

The answer arrived in 2001.

Gleevec: A Milestone that Changed Everything

In chronic myeloid leukemia (CML), scientists discovered that a single genetic abnormality—known as the Philadelphia chromosome—produces a rogue protein (BCR-ABL) that triggers uncontrolled cell division.

Researchers at Novartis developed imatinib (Gleevec), a drug that specifically blocked this protein.

The results were astonishing:

- CML patients who had faced certain death now saw normal life expectancy.
- Side effects were minimal.
- The drug worked like a switch—turning off cancer's fuel line.

It was called "the magic bullet."

And it ushered in the era of targeted therapy.

Precision Follows: A Flood of Smart Weapons

From Gleevec forward, cancer research shifted to biomarkers and molecular profiling. New drugs followed:

Herceptin (trastuzumab) for HER2+ breast cancer

- EGFR inhibitors for certain lung cancers
- ALK inhibitors for rare genetic fusions
- VEGF inhibitors like Avastin to block the blood supply to tumors
- PARP inhibitors for BRCA-mutated ovarian cancers

Each was designed not to destroy everything—but to disrupt one specific pathway.

And with genetic testing, oncologists began tailoring treatment to the individual, not just the disease.

The Human Stories: Survival, Redefined

"I was told I had six months."

That's what Elena, a 42-year-old mother with advanced lung cancer, was told in 2010. But her tumor tested positive for EGFR mutation. She was started on a targeted pill.

"It's been 11 years. I've seen my son graduate from high school. I've hiked mountains. I take one pill a day. That's my chemotherapy."

This is the new narrative:

Live with cancer. Thrive with cancer. Age with cancer.

Beyond Targeted: The Next Frontier

Targeted therapy opened the door to personalized medicine. And now, it walks alongside:

Immunotherapy: drugs that unleash the body's own immune system

- CAR-T cell therapy: engineered immune cells that hunt cancer
- Liquid biopsies: blood tests that detect mutations in real time
- Tumor-agnostic therapies: drugs based on mutation type, not tumor location

What began as a war is now a chess match—strategy, timing, adaptation.

From One Size Fits All to One Patient at a Time

Cancer treatment today is about:

- Knowing the mutation driving the cancer
- Understanding the patient's biology and preferences
- Crafting a plan that balances efficacy and quality of life

We no longer ask, "What chemo do we have?"

We ask, "What makes this tumor tick—and how do we turn it off?"

A Future Full of Second Chances

From pediatric leukemia to metastatic breast cancer, from once-deadly melanomas to hard-to-treat lung cancers, targeted therapy has given millions not just more time, but better time.

" I'm not cured," said one survivor. "But I'm living like I am. And for me, that's enough."

The Revolution isn't Over—It's just getting Smarter

We still lose lives to cancer.

We still face disparities in access.

But we also have something we didn't have before:

Tools. Time. Strategy. Hope.

From the dark days of mustard gas to the bright promise of personalized medicine, this journey reminds us:

We can't always prevent cancer. But we can fight it smarter, gentler, and longer than ever before.

The Story behind Gleevec: Turning Leukemia into a Manageable Disease

How a Deadly Blood Cancer met its Match in a Little Orange Pill that Changed Everything

For decades, chronic myeloid leukemia (CML) was a slow-moving executioner.

Doctors called it a "quiet" cancer—it didn't scream at first.

Patients often felt fine.

Until they weren't.

The average life expectancy after diagnosis?

Three to five years.

Chemotherapy helped for a while.

Bone marrow transplants were possible—but rare, risky, and not widely accessible.

For most, CML was not curable. It was a waiting game.

But then came a breakthrough so unexpected, so precise, and so powerful that it didn't just treat the disease—it redefined cancer care itself.

That breakthrough was called Gleevec.

And it turned a fatal leukemia into a manageable condition with near-normal life expectancy.

The Clue in a Chromosome: Where the Story Begins

In the 1960s, two scientists—Peter Nowell and David Hungerford—noticed something strange in the blood cells of patients with CML:

A tiny, abnormally short chromosome.

Later dubbed the Philadelphia chromosome, this oddity was more than a curiosity. It turned out to be the genetic driver of the disease.

Researchers discovered that it was caused by a translocation—a swapping of genetic material between chromosomes 9 and 22.

This created a fusion gene, called BCR-ABL, which made a rogue enzyme (a tyrosine kinase) that told white blood cells to multiply uncontrollably.

This wasn't just cancer. It was a biochemical mistake—a single switch flipped in the DNA.

If that switch could be turned off, CML might be stopped.

The Scientist who didn't Let Go

In the 1990s, a researcher named Dr. Brian Druker at Oregon Health & Science University became obsessed with the idea:

Could we block that exact enzyme—nothing more, nothing less—and stop the disease at its source?

Working with Novartis and molecular chemists, Druker helped identify a compound—STI571—that could specifically inhibit the BCR-ABL tyrosine kinase.

In petri dishes, CML cells died rapidly while normal cells were left unharmed.

In mice, the drug halted leukemia progression.

It was precise. It was elegant.

And it worked.

They renamed the drug imatinib.

Its commercial name would become legendary: Gleevec.

The Clinical Trials that Changed Lives

The first human trials began in 1998.

Doctors expected modest improvement. What they got was a miracle.

Patients who had run out of options saw their white blood cell counts normalize.

Bone marrow samples showed a dramatic reduction of leukemia cells.

Many went from bedridden to back at work within weeks.

In a landmark Phase II trial, 95% of CML patients went into complete hematologic remission. Some saw their Philadelphia chromosome vanish.

"It felt like watching Lazarus rise," one oncologist said. "Patients who were dying started living again."

FDA Approval in Record Time

In May 2001, the FDA approved Gleevec in record time—just two and a half months after the application was submitted.

It was one of the first drugs ever fast-tracked due to overwhelming efficacy.

And overnight, the world changed for people with CML.

A Pill that Redefined Cancer

Gleevec didn't cure cancer.

- It controlled it.
- Taken once daily
- With minimal side effects

For many patients, this meant normal blood counts, normal energy levels, and a normal life.

It was the first cancer drug designed to target a specific molecular site—not the entire body.

And it proved that cancer could be managed like a chronic illness, not just a ticking clock.

"Before Gleevec, we hoped to extend life by months," Druker said. "Now, we talk about decades."

The Human Impact: Stories of Time Restored

James, diagnosed at 43, was preparing for a transplant he might not survive. His doctor offered him Gleevec instead.

"Within two weeks, I felt better than I had in years. That was 21 years ago."

Lily, a schoolteacher, feared she wouldn't see her daughter graduate. She's now a grandmother.

Thousands of patients, once written off by statistics, were writing their own stories—filled with holidays, anniversaries, promotions, and retirement.

The Ripple Effect: A New Era in Oncology

Gleevec was more than a drug. It was a blueprint:

- Drug development shifted toward targeted therapy
- Research focused on genetic mutations, not just tumor types
- Pharma embraced the "smart drug" revolution

Today, dozens of cancers are treated with therapies inspired by Gleevec's precision—HER2 inhibitors, EGFR blockers, BRAF-targeting agents, and more.

The age of one-size-fits-all chemotherapy began to give way to tailored, tumor-specific strategies.

The Legacy of Gleevec

- Over 300,000 people with CML have been treated with Gleevec or its next-generation versions.
- The 10-year survival rate for CML is now over 85%—an unimaginable achievement before the introduction of Gleevec.
- And most of those patients are living full, active, productive lives.

"It was a pill," one survivor said. "But to me, it was a passport back to life."

Gleevec didn't just save lives.

It changed what survival looked like.

It told the world: cancer doesn't always have to win.

And that some battles can be won quietly, one pill at a time.

Survivor Stories: " I Lived to See My Granddaughter"

Survivor Stories from the Frontlines of Modern Medicine

When Meera, 59, was diagnosed with chronic myeloid leukemia (CML) in 2003, her world shrank in an instant.

She had gone to her doctor for fatigue. She was still working, planning a vacation, helping her daughter prepare for marriage. She had dreams of retirement—of learning to paint, of someday becoming a grandmother.

Instead, she was told she had a rare blood cancer. One that had no cure. One that used to be fatal.

She remembers hearing the word leukemia and feeling like the walls of the room were closing in.

"I couldn't think. I couldn't speak. All I could see was that I might not be there when my daughter had a baby."

But what Meera didn't know then was that she was diagnosed in a new era—a time when medicine was no longer relying only on chemo and hope.

She was about to meet the drug that would give her back her future.

A Small Pill, A Giant Hope

Her oncologist told her about Gleevec, a daily tablet that blocked the abnormal enzyme causing her leukemia.

"It's not a cure," he said. "But for many people, it works like one."

Meera started taking the little orange pill the next day. She expected harsh side effects. She expected to be bedridden. Instead, something remarkable happened.

Within a few weeks, her energy returned.

Within a few months, her blood counts returned to normal.

Within a year, her cancer was in complete molecular remission.

"I still had cancer," she said, "but it didn't have me."

The Days she thought she'd never See

In 2006, Meera stood beside her daughter at her wedding.

In 2009, she retired from her government job and began teaching neighborhood children how to draw.

In 2012, her daughter called with tears in her voice and joy in her heart:

"Amma, you're going to be a grandmother."

That day, Meera held her Gleevec bottle in her hand and whispered, "Thank you."

Holding the Future in her Arms

When baby Anika was born, Meera was in the delivery room.

"She opened her eyes and grabbed my finger. I thought—I wasn't supposed to be here. But here I am."

She watched her take her first steps, say her first words, and celebrate her first birthday.

Every photo on Meera's phone was a moment medicine made possible.

More than Survival—Life Fully Lived

Today, Meera:

- Takes a walk every morning
- Keeps a journal of all the milestones she's witnessed
- Leads a support group for women with cancer diagnoses in her town
- Sends her oncologist a birthday card every year

"I'm not a miracle," she says. "I'm a reminder of what science can do when it listens, when it cares."

Echoes of her Story around the World

Meera's journey is no longer unique. It's part of a growing chorus of voices made possible by breakthroughs like Gleevec, Herceptin, Keytruda, and many others:

- A lung cancer survivor who lived to walk his daughter down the aisle.
- A breast cancer survivor who held her grandson while still on maintenance therapy.

- A leukemia survivor who celebrated his 80th birthday with four generations around him.

"It's not just about living longer," Meera says. "It's about living well enough to show up for the people you love. That's the gift."

The Legacy of a Second Chance

Meera once thought her story would end quietly and quickly.

Instead, it continues—loudly, joyfully, with finger paints, fairy tales, and a little girl who calls her 'Amma.'

She doesn't know what tomorrow holds.

But she knows what today means.

" I lived to see my granddaughter," she says. "Everything after that is a bonus."

Chapter 16

Vaccines: Saving Millions Quietly

The Unseen Soldiers that Changed the Course of Human Life

There are no parades for diseases that never happened.

No headlines for the child who didn't die of measles.

No breaking news for the mother who didn't pass tetanus to her newborn.

No front-page story for the village where nobody got polio this year.

And yet, quietly, consistently, and almost invisibly, vaccines have saved more lives than perhaps any other medical intervention in history.

They don't heal the sick.

They prevent sickness in the first place.

They don't work by drama—but by disruption.

They cut off illness before it begins, and in doing so, change the trajectory of families, generations, and nations.

Vaccines are the quiet heroes of medicine.

A World before Vaccines

It's easy to forget what life once looked like:

- Smallpox killed 3 out of every 10 people it infected.
- Polio left children paralyzed and in iron lungs.
- Measles caused blindness, encephalitis, and death in hundreds of thousands.
- Tetanus, diphtheria, whooping cough—they were feared words, not chapter titles in a history book.

Parents buried their children. Communities shut down during outbreaks.

A cough could mean catastrophe. A fever could be final.

Then came a revolution that didn't cure disease—it stopped it from starting.

The First Breakthrough: Smallpox and the Birth of Modern Immunization

In 1796, Edward Jenner scratched cowpox pus into a boy's arm and discovered immunity to smallpox.

It was radical. Unethical by modern standards. But it worked.

More than 150 years later, mass vaccination campaigns eliminated smallpox entirely. In 1980, the World Health Organization declared the disease eradicated.

It remains the only human disease we've completely removed from the planet—thanks to a vaccine.

The 20th Century: A Golden Age of Vaccines

Between the 1920s and 1980s, vaccines were developed for:

- Diphtheria
- Pertussis (whooping cough)
- Tetanus
- Polio
- Measles, mumps, rubella (MMR)
- Hepatitis B

These vaccines didn't just change health—they changed childhood.

Where once parents braced for loss, they now expected laughter.

Where once epidemics closed schools, students now planned their futures.

Polio: A Battle Nearly Won

In the 1950s, polio paralyzed thousands every year.

When Jonas Salk's vaccine arrived, families lined up in droves.

Later, Albert Sabin's oral vaccine made mass immunization easier globally.

Today, polio is nearly eradicated, with only 0.1% remaining.

Just a few pockets remain—and global health leaders continue to chase it to the finish line.

"When polio is finally gone," said one health worker, "it will be the quietest global miracle in history."

Hepatitis B and Cervical Cancer: The Hidden Wins

Some vaccines fight not just viruses—but cancer.

- The hepatitis B vaccine drastically reduces the risk of liver cancer
- HPV vaccine prevents strains of human papillomavirus that cause cervical, throat, and anal cancers

Girls and boys vaccinated today may never know they were spared a life-threatening illness.

And that's the beauty of it: they'll never know. And they won't have to.

The Math of Miracles

According to the WHO:

- Vaccines prevent 4 to 5 million deaths every year
- For every $1 spent on childhood immunizations, the return in societal benefits is $44
- Diseases once widespread are now rare or forgotten in many countries

But make no mistake: the success of vaccines is not that we see it—

It's that we don't.

The COVID-19 Vaccine: Science at Breakneck Speed

In 2020, the world faced a new enemy.

Less than a year after SARS-CoV-2 emerged, vaccines were approved—faster than any in history.

mRNA platforms, such as Pfizer-BioNTech and Moderna, became the frontline.

Billions of doses were delivered in record time.

Hospitals emptied. Death rates dropped. Economies reopened.

It was a triumph not just of science, but of decades of vaccine research quietly waiting in the wings.

The Resistance and the Reminder

Despite their impact, vaccines have faced misinformation, fear, and hesitancy.

But their value remains unchanged.

They are:

- Safe
- Proven
- One of the most effective public health tools ever created

"Vaccines don't make headlines," one physician said. "They make birthdays."

Quiet Medicine, Loud Impact

A vaccine never says, "You've been cured."

It says: "You were never sick to begin with."

It spares the funeral.

It saves the memory.

It preserves the laugh of a child who never got sick enough to be quieted.

The Legacy of the Unseen

Millions will grow up never knowing the pain their parents feared.

Millions more will age, work, travel, and thrive—because something silent and powerful stood guard in their blood.

That is the story of vaccines.

They save lives not with fanfare, but with frequency.

Not with noise, but with numbers too large to ignore.

"I don't remember getting vaccinated," said a survivor of the measles era. "But I'll never forget what I saw before vaccines came."

Hepatitis B, HPV, Rotavirus, and Beyond

The Silent Killers that Vaccines Stopped and the Futures they Quietly Saved

Some diseases don't arrive with a dramatic flair.

They don't paralyze or disfigure. They don't come with headline-grabbing outbreaks.

They move quietly, relentlessly, and often without symptoms—until it's too late.

But thanks to vaccines, millions of people are now spared from the cancers, infections, and lifelong complications these diseases once caused.

This chapter tells the story of three such vaccines—Hepatitis B, HPV, and Rotavirus—and how they changed the odds for entire generations.

They don't get the glory. But they deserve it.

Hepatitis B: The Liver's Lifesaving Shield

"He didn't look sick… until he was dying."

Hepatitis B is a viral infection that attacks the liver, spreading through blood and bodily fluids. It often causes no symptoms for years—until it suddenly manifests as liver failure or cancer.

It is:

- One of the most common infections worldwide, especially in Asia and sub-Saharan Africa
- A leading cause of liver cancer
- Transmissible from mother to child at birth, silently passed from one generation to the next

In the 1980s, a vaccine was developed—one of the first to prevent a type of cancer.

Today:

- Hepatitis B vaccination is part of the standard infant immunization schedule in over 180 countries

- In countries with high coverage, liver cancer rates in children have plummeted
- Perinatal transmission has dramatically declined

"My son was born with Hep B," said one mother in Taiwan. "My grandson wasn't. That vaccine broke the chain."

HPV: A Cancer Vaccine in Disguise

"She was only 32 when she was diagnosed with cervical cancer."

Human papillomavirus (HPV) is the most common sexually transmitted infection in the world.

Most infections clear naturally. But some strains—especially HPV-16 and HPV-18—can cause cervical, throat, anal, and penile cancers.

Cervical cancer kills over 300,000 women a year—mostly in low-income countries.

Then came a turning point.

In 2006, the first HPV vaccine (Gardasil) was approved.

It targeted the strains most likely to cause cancer.

Now, the science is staggering:

- In countries with high HPV vaccine coverage, cervical cancer rates in young women have dropped by up to 90%
- Australia is on track to eliminate cervical cancer as a public health problem within decades
- Vaccination of boys and men has expanded protection and reduced transmission

"My daughter will never know what it's like to fear a Pap smear," said one survivor. "That's the gift the vaccine gave her."

Rotavirus: Ending the Cycle of Infant Diarrheal Deaths

"It wasn't the fever that scared me—it was how quickly he became weak."

Rotavirus doesn't sound terrifying.

But before the vaccine, it was the leading cause of severe diarrhea and dehydration in infants and young children.

In developing countries, where clean water and IV access were limited, rotavirus killed over 500,000 children a year.

It struck fast, often in the first year of life. A baby who was smiling in the morning could be unresponsive by night.

Two vaccines—Rotarix and RotaTeq—Changed Everything

Since their introduction:

- Hospitalizations for diarrhea in children have dropped by 80–90% in many countries
- Mortality from rotavirus has plummeted globally
- Parents no longer fear a simple stomach bug turning deadly

"I've seen what rotavirus does to babies," said a pediatrician in rural India. "Now I barely see it at all."

Beyond These: A Quiet Revolution Expands

While Hep B, HPV, and rotavirus transformed care for specific diseases, they are part of a larger shift in global vaccination strategy:

- The pneumococcal vaccine now prevents deadly pneumonia and meningitis in children.
- The Haemophilus influenzae type b (Hib) vaccine has nearly wiped out a once-common cause of infant meningitis.
- Meningococcal vaccines prevent sudden, deadly brain infections in adolescents and college-age adults.
- Japanese encephalitis and yellow fever vaccines protect millions in endemic regions.
- Typhoid conjugate vaccines are now being administered to children in South Asia and Africa.

Each one tells the same story:

A child is protected. A parent spared grief. A future made possible.

Vaccines as Equity, Not just Medicine

These vaccines don't just protect the body. They protect:

- Economies: fewer hospitalizations, more children in school, more productive adults
- Health systems: reduced burden on fragile infrastructure

- Families: fewer funerals, more celebrations

They represent health justice—especially when extended to the poorest corners of the world through initiatives like Gavi, the Vaccine Alliance, and UNICEF partnerships.

The Unseen Impact

You rarely see a rotavirus survivor holding a sign.

You won't meet many people who say, "HPV almost got me."

And the child who didn't get hepatitis B at birth? They'll never know the future they narrowly escaped.

And that's the point.

"When a vaccine works," one doctor said, "nothing happens. And that's everything."

A Legacy that Lives Quietly

Every vial, every drop, every oral dose saves a life—or a lifetime of suffering.

They're not dramatic. But they are profound.

They are:

- The difference between illness and innocence
- Between cancer and career
- Between a grieving parent and a grateful one

"I don't remember my daughter's vaccinations," one mother said.

"I just know she's healthy. And that means someone, somewhere, did their job right."

Voices from the Field: The Global Health Workers' Crusade

The People who Walked Miles so Others could Live

They don't wear white coats or appear on the covers of scientific journals.

They're not celebrated in Nobel speeches.

But they are the beating heart of public health.

They are nurses, midwives, community health workers, logistics teams, and volunteers.

The ones who walk for hours with cold boxes on their backs.

The ones who hold down crying children with one hand and comfort mothers with the other.

The ones who bring life-saving care to places maps barely reach.

They are the invisible army in the war against disease.

This chapter is their story.

"I Carry a Cold Box, But What I'm Really Carrying Is Trust."

Fatima, a community health worker in northern Nigeria, has been vaccinating children against polio for over a decade.

She walks village to village under searing heat, navigating terrain where cars can't go.

"At first, people didn't trust us. Some thought the vaccines made children infertile. Some locked their doors. But we came back. Every week. Every child. Until they saw we weren't bringing harm—we were bringing hope."

Fatima has vaccinated thousands.

She's watched paralyzed children who arrived too late.

And she's celebrated every year without a new case of polio.

"When we hear there are no more cases, we dance. Quietly, respectfully—but we dance."

"I Lost My Brother to Measles. That's why I do this Work."

Joseph, a health volunteer in rural Uganda, still remembers the night his younger brother died.

"He was only four. A fever came. Then convulsions. Then he stopped breathing. It was measles. We didn't even know it had reached our village."

Now, Joseph leads immunization outreach.

He speaks in markets, schools, churches.

He carries coolers packed with rotavirus and MMR vaccines across rivers and hills.

"Every time I vaccinate a child, I think of my brother. He died because no one reached us in time. I want to be the reason someone else doesn't say the same."

"We Go where the Diseases Go."

Anita, a nurse in Bangladesh, has worked through cyclones, floods, and cholera outbreaks.

"We vaccinate in disaster shelters. We treat in waist-deep water. We chase cholera, typhoid, and hepatitis A like firefighters chasing smoke."

She recalls carrying oral cholera vaccine through knee-deep mud during monsoon season.

"People ask why we do this. I say—because no one else will."

"I'm Not just giving a Shot. I'm giving a Promise."

Tenzing, a health worker in the Himalayas, rides a yak part of the way to reach remote villages.

He's vaccinated infants against hepatitis B at altitudes where oxygen is thin and roads disappear.

"The mothers wait for us, wrapped in blankets, babies in arms. They've walked for hours, too. I look at them and think—we're meeting in the middle, for the same reason: to protect life."

The Daily Realities: Hard Roads, Harder Choices

Health workers around the world face:

- Long distances—often walked without pay

- Hostility and violence in conflict zones
- Lack of resources, cold chain failures, and political resistance
- Tremendous emotional burden—especially when children die in their arms

Yet they return.

Because they know that every vaccine delivered is a life saved.

Every trust built is a barrier against misinformation.

Every smile from a mother is a sign the crusade is working.

When COVID Came, they didn't Step Back—they Stepped Up

During the COVID-19 pandemic, these same workers:

- Delivered PPE to rural clinics
- Carried vaccines to the last mile
- Educated communities battling fear and conspiracy
- Held the hands of the dying—and the newly recovered

"We didn't get applause," said a midwife in the Philippines. "But we got results. That's better."

Their Legacy is Measured in Children who Grow Up

Global health workers don't ask for praise.

They ask for:

- Full vaccine funding
- Safety and fair pay
- Training and respect
- A system that supports their service

They don't see themselves as heroes.

But every mother who didn't have to bury a child, every village that no longer fears polio, every baby born into a world just a little safer—tells a different story.

"We are not the face of science," said Fatima. "But we are the hands and the feet. We walk science into the lives that need it."

The Final Mile is Never Easy—But it's always Worth it

This is not just a job. It's a crusade.

A fight for dignity, equity, and survival.

A belief that no child, anywhere on Earth, should die of a disease we know how to prevent.

And because of them—the quiet warriors in worn shoes and sun-scorched clothes—the world has come closer to that dream than ever before.

PART V: THE ERA OF BIOLOGICS AND PERSONALIZED MEDICINE 2000s – 2020s)

Part V

The Era of Biologics and Personalized Medicine (2000s - 2020s)

How Medicine Moved from Mass Prescription to Tailored Precision—and Patients became the Center of the Story

For much of medical history, treatment was a numbers game.

Doctors used what worked for most people, most of the time, hoping their patient wasn't the exception.

Prescriptions were written according to protocol, not by individual physicians.

Then, the 21st century arrived—and medicine underwent a significant shift.

A quiet revolution began—one that didn't just target diseases but targeted patients.

A new era was born, shaped by two powerful forces:

- Biologics: precision-engineered therapies made from living cells
- Personalized medicine: the tailoring of treatment based on your unique biology

Together, they marked a turning point where science no longer asked, "What disease is this?" but rather,

"Who is this patient—and what is their disease made of?"

Biologics: Medicines Made from Life Itself

Unlike traditional small-molecule drugs (aspirin, statins, chemotherapy), biologics are:

- Produced using living organisms (like yeast or mammalian cells)
- Often proteins, antibodies, or hormones
- Larger, more complex, and exquisitely targeted

The most famous early biologic? Insulin—but what followed reshaped modern medicine.

Biologics revolutionized care for diseases once considered untreatable, including:

- Autoimmune conditions like rheumatoid arthritis, psoriasis, and Crohn's disease
- Cancers, especially breast, colon, and blood cancers
- Rare genetic disorders
- Asthma and severe allergies
- Multiple sclerosis and migraines

Blockbuster biologics like:

- Adalimumab (Humira)
- Trastuzumab (Herceptin)
- Etanercept (Enbrel)
- Pembrolizumab (Keytruda)
- didn't just suppress symptoms—they altered disease courses, improving outcomes and quality of life.

"It was like flipping a switch," said one Crohn's disease patient. "I went from planning my life around a bathroom to planning a vacation."

The Rise of Monoclonal Antibodies: Smart Weapons

Many biologics are monoclonal antibodies—engineered molecules that:

- Bind to specific targets on cells, like cancer or immune cells
- Block rogue signals
- Flag invaders for immune destruction

Instead of hitting the whole body, these drugs act with laser precision.

For example:

- Herceptin binds only to HER2-positive breast cancer cells
- Rituximab targets B-cells in certain leukemias and lymphomas
- Dupilumab neutralizes inflammatory pathways in eczema and asthma

"We've moved from 'kill everything' to 'strike surgically', said one oncologist.

Personalized Medicine: When Treatment Fits the Patient

While biologics reshaped the tools, personalized medicine reshaped the mindset.

Driven by genomics, molecular diagnostics, and data science, personalized medicine asks:

- What genes does this tumor carry?
- What's the patient's metabolism like?
- What drugs will work best for them—and only them?

Examples include:

- Oncotype DX testing to determine if early breast cancer needs chemo
- BRCA testing to guide cancer prevention in high-risk individuals
- EGFR and ALK mutation testing in lung cancer to guide therapy
- Pharmacogenomics to determine how patients metabolize drugs like warfarin or antidepressants

The Impact on Real Lives

Sophie, diagnosed with breast cancer at 39, was terrified of chemotherapy.

Her oncologist ran a genomic risk score. The results showed no benefit from chemo—only hormone therapy was needed.

"The test didn't just spare me the side effects. It gave me confidence that I was doing the right thing for my body."

Tariq, with rheumatoid arthritis, tried multiple drugs with no success. A genetic biomarker flagged IL-6 pathway dysfunction. A specific biologic worked—and within weeks, his joint pain subsided.

From the Genome to the Pharmacy

The Human Genome Project was completed in 2003.

By the 2020s, DNA sequencing had become faster, cheaper, and clinically actionable.

Today:

- Your tumor can be sequenced to guide treatment
- Your genes can inform your drug doses
- Your immune system profile can predict your response to therapy

It's no longer "one-size-fits-all."

It's "this-size-fits-you."

COVID-19 and the Biologic Test of Fire

The pandemic accelerated everything:

- mRNA vaccines (Pfizer and Moderna) became the first widely used biologics produced at unprecedented speed
- Monoclonal antibody therapies helped high-risk patients
- Genomic surveillance tracked viral mutations in real time

COVID didn't just challenge medicine—it validated the biologics and personalized medicine playbook.

Challenges alongside Breakthroughs

This revolution brought new questions:

- Access and affordability: Biologics are often expensive
- Equity: Are personalized tools available only to the privileged?
- Data privacy: Who owns your genetic information?
- Complexity: How do we interpret a flood of molecular data?

But these are growing pains—signs of progress, not regression.

A Future where You are the Formula

In this era, you're not just a diagnosis.

You are a blueprint.

A set of molecular signals, immune patterns, genetic risks, and possibilities.

"We used to treat the average patient," one physician said.

"Now, we treat you."

The Era that Quietly Changed Everything

The years between 2000 and 2020 didn't bring a cure for every disease.

But they changed the questions we ask.

They changed the tools we use.

And they changed the way patients see themselves:

Not as passive recipients of treatment, but as partners in precision healing.

Chapter 17

Monoclonal Antibodies: New Era in Cancer and Autoimmune Diseases

How Lab-Made Precision Tools Turned the Immune System into a Life-Saving Ally

Once upon a time, the immune system was considered a mystery—brilliant but blunt, powerful but imprecise.

It could kill viruses, protect against infections, and even reject organs. But it was unpredictable, untrainable, and often uncontrollable.

Then Science did something Extraordinary

It learned how to make artificial antibodies—molecules that could be programmed to target one thing, and only that thing, with surgical precision.

This is the story of monoclonal antibodies (mAbs)—the lab-born molecules that changed how we fight cancer, autoimmune disorders, and chronic disease.

They are not just drugs.

They are guided missiles.

And their arrival marked the birth of intelligent immunology.

What are Monoclonal Antibodies?

Monoclonal antibodies are:

- Engineered versions of the proteins your immune system already makes
- Designed to bind to a specific target—like a protein on a cancer cell or an overactive immune molecule
- Mass-produced from a single immune cell clone (hence mono-clonal)

Imagine them as custom-designed keys—each made to fit one molecular lock.

And when they turn that lock, they can:

- Block signals (e.g., cancer growth signals)
- Mark cells for destruction by the immune system
- Interrupt inflammation in autoimmune diseases
- Carry drugs or toxins directly to diseased cells

The First Breakthrough: Cancer Meets its Match

The earliest success stories came in oncology.

Rituximab (1997)

Targeted CD20 on B-cells, revolutionizing the treatment of non-Hodgkin lymphoma and later rheumatoid arthritis.

"Patients who were dying started walking out of the hospital," said one oncologist. "We hadn't seen anything like it."

Trastuzumab (Herceptin, 1998)

Targeted HER2 receptors in aggressive breast cancer, slashing recurrence and improving survival.

"It turned a frightening diagnosis into something we could fight—and win," said a survivor.

Bevacizumab (Avastin, 2004)

Targeted VEGF to cut off the blood supply to tumors in colon, lung, and brain cancers.

Then came Autoimmune Diseases

Inflammatory diseases like rheumatoid arthritis, Crohn's disease, psoriasis, and ulcerative colitis were once managed with steroids and generalized immunosuppression.

Then came monoclonal antibodies that:

- Blocked TNF-α (e.g., infliximab, adalimumab)
- Inhibited IL-6, IL-17, IL-23
- Targeted B-cells, T-cell co-stimulation, or IgE antibodies

Suddenly, patients who had been bedridden could:

- Work again
- Travel again
- Hold their children again
- Plan futures instead of pain management

Neurology, Asthma, and More

The mAb revolution spilled into other areas:

- MS (multiple sclerosis): Natalizumab and ocrelizumab stopped disease progression
- Asthma: Dupilumab and omalizumab reduced attacks in severe cases
- Migraines: CGRP-targeting antibodies brought relief after decades of suffering
- Allergies: mAbs are helping desensitize patients to life-threatening allergens

"One injection changed my year," said a chronic migraine patient. "And gave me back my brain."

How they Work: Four Main Modes

1. Target and destroy: e.g., binding to cancer cells and marking them for immune attack
2. Block bad signaling: e.g., intercepting growth factors in tumors
3. Deliver payloads: e.g., carrying chemotherapy or radioactive particles directly to diseased tissue
4. Calm the immune system: e.g., suppressing overactive inflammation in autoimmune disease

They're not sledgehammers.

They're scalpels.

The Patient Experience: A New Kind of Medicine

Most mAbs are:

- Given via infusion or injection
- Needed every few weeks or months
- Come with fewer side effects than older drugs
- Require monitoring, but offer life-changing relief

Patients often describe starting mAbs as the moment they started living again.

"I was on 12 pills a day. Now, it's one infusion a month—and I can pick up my grandkids," said a man with psoriatic arthritis.

Challenges and Cautions

While revolutionary, mAbs come with:

- High cost (often tens of thousands per year)

- Immune-related side effects (e.g., infusion reactions, infections)
- Complex manufacturing and delivery logistics
- Limited access in low-income countries

But with the rise of biosimilars—generic versions of biologics—costs are coming down, and access is expanding.

Legacy and Future

Today, over 100 monoclonal antibodies are approved for human use.

They've helped:

- Cure some leukemias
- Tame lifelong autoimmune conditions
- Save lives during Ebola and COVID outbreaks
- Bring precision where there was once only hope

And the future?

It includes bispecific antibodies, antibody-drug conjugates, and off-the-shelf cancer immunotherapies.

A Quiet Revolution with Loud Results

Monoclonal antibodies don't make miracles overnight.

But they rewrite biology in real time.

They teach the immune system how to see again.

How to fight smarter.

How to heal without harming everything else.

"These aren't just medicines," said one immunologist.

"They're proof that we can design healing—one molecule at a time."

The Birth of Rituximab, Herceptin, and Keytruda

How Three Precision Drugs Changed the Way we Fight Cancer—and gave Millions a Second Chance

For decades, cancer treatment was a war waged with few weapons and even fewer rules.

Chemotherapy didn't ask where the enemy was—it carpet-bombed the body, hoping cancer cells would die faster than healthy ones.

Then came a new approach: targeted therapy.

Drugs that didn't attack everything, but rather something specific, vital, and vulnerable in cancer cells.

This chapter tells the story of three such drugs—Rituximab, Herceptin, and Keytruda—and the people whose lives they changed.

Each was a breakthrough.

Each gave rise to a new way of thinking.

Each was born from vision, doubt, and dogged belief that cancer could be stopped more cleanly, more kindly, and more personally.

Rituximab: The First Monoclonal Antibody Approved for Cancer

"We didn't think the body could fight back. Then Rituximab showed us it could."

In the 1980s, scientists were racing to harness monoclonal antibodies—lab-made proteins that could target specific molecules.

Researchers at IDEC Pharmaceuticals, a small biotech company, discovered that in non-Hodgkin lymphoma, a protein called CD20 is expressed on the surface of malignant B-cells.

They asked: What if we designed an antibody that stuck to CD20 like glue—and let the immune system do the rest?

They did. And they named it Rituximab.

When tested in patients, something remarkable happened:

- Tumors shrank rapidly
- Side effects were far less severe than chemo
- The immune system wasn't destroyed—it was enlisted

In 1997, Rituximab became the first monoclonal antibody approved for cancer.

It changed the standard of care overnight—not just for lymphoma, but for autoimmune diseases like rheumatoid arthritis and lupus years later.

"Before Rituximab, remission was rare. After Rituximab, it became routine," said one oncologist.

Herceptin: Cracking the Code of Breast Cancer

"They told me I had a kind of breast cancer that was aggressive. They didn't tell me there was a drug that spoke its language."

In the early 1980s, breast cancer was considered a single disease. But researcher Dr. Dennis Slamon at UCLA discovered something shocking:

30% of breast cancer tumors carried extra copies of a gene called HER2, which made them grow faster, spread quicker, and return more often.

It wasn't a death sentence—it was a different disease entirely.

Dr. Slamon worked with Genentech, where scientists developed trastuzumab, a monoclonal antibody that targeted HER2 specifically.

It was risky. Trials were underfunded. Support was thin.

But patients—especially women with no other options—enrolled. And when the results came back, they weren't just hopeful—they were historic.

- Tumor growth slowed dramatically.
- Disease-free survival improved
- HER2+ cancer, once feared, became treatable, manageable, and survivable

In 1998, Herceptin was approved.

"It didn't just save lives," said Dr. Slamon. "It proved that understanding a tumor's biology wasn't academic—it was everything."

Keytruda: Teaching the Immune System to See the Enemy

"What if the body already had the answer—and we just needed to unlock it?"

In the early 2010s, a new approach to cancer treatment, immunotherapy, was gaining attention.

Scientists had discovered that tumors evade the immune system through a protein called PD-1.

By switching it "on," they sent a false signal: "I'm part of you. Don't attack."

What if we blocked that signal?

Enter Keytruda (pembrolizumab)—a drug developed by Merck that inhibited PD-1, freeing the immune system to recognize and attack the tumor.

In 2013, a melanoma patient named Jimmy Carter was given Keytruda for brain metastases.

He recovered.

He lived.

And the world took notice.

Keytruda wasn't just for melanoma—it began working for lung cancer, bladder cancer, head and neck cancer, Hodgkin's lymphoma, and more.

In 2017, Keytruda became the first drug ever approved based on a tumor's genetic feature (MSI-high)—not its location in the body.

"It wasn't just a drug," said one researcher. "It was the start of a new paradigm: Tumor-agnostic therapy."

Three Drugs, One Message: Precision Saves

Together, Rituximab, Herceptin, and Keytruda:

- Redefined cancer biology
- Recruited the immune system as an ally
- Turned once-lethal diagnoses into manageable, chronic conditions
- Sparked a global wave of biologic innovation that still accelerates today

The Human Side: Lives that Weren't Supposed to Last

Lena, diagnosed with HER2+ breast cancer at 34, was told her odds weren't good. She was given Herceptin.

Ten years later, she volunteers at the same hospital where she was once a patient.

Marco, a veteran with lymphoma, failed chemo three times. Rituximab worked. He walked his daughter down the aisle.

Jasmine, diagnosed with metastatic lung cancer in her 50s, was placed on Keytruda.

Three years later, she hikes mountains with her grandchildren.

"These drugs didn't just extend our lives," Jasmine says. "They returned them to us."

Legacy: The Door they Opened

These three biologics changed:

- How we classify cancers
- How we design drugs
- How we understand immune response
- How we hope

They proved that cancer care didn't have to be crude, cruel, or uniform.

It could be thoughtful. Personal. Strategic. And stunningly effective.

"These drugs taught us something profound," said one researcher.

"You don't always need to fight harder. You need to fight smarter."

Patient Chronicles: "The Tumor Shrunk, I Got My Life Back."

How a New Generation of Therapies didn't just Shrink Disease—but Rebuilt Futures, One Life at a Time

Sonia was 42 when she was diagnosed with stage IV non-small cell lung cancer.

She had never smoked. She ran 10Ks. She had two kids in school and a startup just breaking even. The cough she thought was from stress turned out to be a tumor in her left lung—with small metastases to her bones.

The doctor looked at her gently and said, "This isn't curable. But it's treatable."

She heard the word terminal.

But what she didn't yet know was that she was living in a new era—a time when cancer care had moved from blunt to brilliant.

And her body carried a secret weapon: a genetic mutation called EGFR.

That changed everything.

Not a Miracle—A Match

A sample of Sonia's tumor was sent for molecular profiling.

The results came back: she was a candidate for a targeted therapy—osimertinib, a pill that inhibits the abnormal EGFR protein driving her cancer.

No infusions.

No hair loss.

No endless days in a hospital chair.

Just one daily pill.

And a follow-up scan a month later.

That's when her doctor walked in, smiling.

"The tumor shrank."

Not completely. Not cured.

But shrunk enough for her breathing to improve. Her bone pain to lessen. Her voice to regain strength. Her mind to imagine next week. Next year.

The Return to Life

Within three months, Sonia was:

- Driving her kids to school again
- Back at work, part-time
- Teaching yoga in her living room
- Planning a vacation—not for escape, but for celebration

She still lived with cancer.

But now, cancer lived with her.

It was something she treated. Not something that defined her.

A Thousand Stories Like Hers

Across the world, modern cancer therapy—biologics, immunotherapy, and targeted treatments—has sparked a revolution in quiet recoveries.

People like:

- Victor, 67, was told his lymphoma was aggressive. He started Rituximab and walked out of the hospital two weeks later feeling normal.
- Lena, 35, whose HER2+ breast cancer was stopped in its tracks with Herceptin. Today, she runs a mentorship group for cancer survivors.
- Jamil, 59, who failed two lines of chemo. Then came Keytruda—an immunotherapy that melted his melanoma away and gave him five more years and counting.

"We were told to get our affairs in order," Lena said.

"Now we're planning holidays and home renovations."

What it Means to Shrink a Tumor—Beyond the Scan

When a tumor shrinks, so does:

- The pain
- The fear
- The sense that life has become a countdown

And what grows is:

- Possibility
- Energy
- Identity beyond illness

"People think the biggest gift was the tumor shrinking," Sonia said.

"But actually, it was getting my ordinary life back. Picking up groceries. Laughing at something stupid. Sleeping without fear."

Not Every Case Ends in Remission—But Many begin Again

These therapies don't always cure. But they often:

- Turn fatal cancers into chronic illnesses
- Delay progression for years
- Offer dignity, comfort, and strength in time that once didn't exist

The tumor shrinks.

The lungs expand.

The room brightens.

The life resumes.

Science behind the Scenes

Sonia's pill wasn't a miracle. It was:

- Years of genomic research
- Clinical trials run across the globe
- A pharmaceutical company that kept investing in lung cancer when others gave up
- A health system that approved biomarker testing before treatment began

It was evidence of meeting a body at the right time.

The New Language of Survival

In this new world, people say:

- "I'm in molecular remission."
- "My scans are stable."
- "I respond to immunotherapy."
- "I take a pill. That's my chemo."
- "The tumor shrank. I got my life back."

The Journey Continues

Sonia still takes her pill every morning with tea.

Every scan brings anxiety. Every good result brings breath.

She's not cured.

But she is living—with intention, with gratitude, and with a future that once disappeared overnight.

"My cancer is still there. But so am I."

Chapter 18

Hepatitis C Cure: From Lifelong Disease to Total Eradication

The Story of a Silent Killer that Science Finally Silenced

For decades, Hepatitis C was one of the world's quietest epidemics.

It didn't bring fevers or rashes.

It didn't travel in outbreaks.

It lived silently in the liver, often for decades—until it triggered cirrhosis, liver failure, or cancer.

And for those who carried it, the prognosis was bleak:

- There was no cure.
- The treatments were brutal and only partially effective.
- And most infected people didn't even know—until their liver began to fail.

But today, this story reads differently.

Because Hepatitis C is now one of the only chronic viral infections that can be completely cured—with pills, in weeks, with minimal side effects.

This is the story of how medicine didn't just treat Hep C—it outsmarted and dismantled it.

The Early Days: A Mysterious Virus with No Name

In the 1970s and 80s, doctors noticed a strange pattern.

Patients receiving blood transfusions developed chronic liver inflammation, but tested negative for hepatitis A and B.

They called it non-A, non-B hepatitis.

It wasn't until 1989 that scientists identified the virus behind it:

Hepatitis C Virus (HCV)—a blood-borne RNA virus that could stay dormant for years, even decades.

By then, millions had already been infected:

- Through transfusions before widespread screening
- Through shared needles
- Through unsafe medical practices
- Sometimes through unknown sources

The First Generation of Treatments: Hard Pills to Swallow

For years, the only option was interferon-based therapy—often paired with ribavirin.

These treatments were:

- Long (6–12 months)
- Harsh (flu-like symptoms, anemia, depression)
- Limited in effectiveness—with only 30–50% cure rates, depending on the virus genotype

Patients feared the cure as much as the disease.

"I tried it. I lost weight. I lost my job. I still had the virus," said one survivor.

The virus was resilient. It hid well. And unlike hepatitis B, it offered no protective immunity—you could get reinfected.

Then Everything Changed: Direct-Acting Antivirals (DAAs)

In 2013, a new class of drugs emerged:

Direct-acting antivirals (DAAs).

These were:

- Oral pills—no injections
- Well-tolerated—few side effects
- Short-course—typically 8 to 12 weeks
- Effective—with cure rates exceeding 95%

The first approved blockbuster was Sofosbuvir (Sovaldi), developed by Gilead Sciences.

It targeted a key enzyme the virus needed to replicate, and when paired with other DAAs, it eradicated the virus completely.

No virus. No relapse. Cured.

The Word we Never Used: "Cure"

In medicine, especially virology, "cure" is a rare and sacred word.

We don't cure HIV. We don't cure most chronic viruses.

But with Hep C, something different happened:

- Patients completed a short course
- Follow-up testing showed no viral RNA
- Even years later, the virus never came back

This wasn't remission.

This was a biochemical victory.

"I was cured," said Mike, a veteran infected decades earlier. "I lived with this for 25 years. Then it was gone—in 12 weeks."

The Human Impact: A Second Life Begins

- Patients awaiting liver transplants came off the list
- Fatigue, pain, and depression lifted after years
- Fear of transmitting the virus to partners or children disappeared
- People who once lived in secret with shame and stigma found freedom

"I had given up on being cured," said Rosa, a former nurse. "Now I dream again."

From Treatment to Elimination

The World Health Organization set an ambitious goal:

Eliminate Hepatitis C as a global health threat by 2030.

The strategy?

- Screening: especially in high-risk populations
- Affordable access to DAAs, including through generic licensing
- Integration into primary care and harm reduction services
- Global partnerships with governments, NGOs, and pharma

Some countries are already leading the way:

- Egypt screened 60 million people and treated over 4 million
- Georgia launched a national Hep C elimination program in 2015
- Iceland is close to zero new cases

Challenges Remain—but So does the Opportunity

Even with a cure in hand, barriers persist:

- Underdiagnosis: most infected people don't know
- Cost: though falling, DAAs remain expensive in some regions
- Access for marginalized groups: including people who inject drugs, prisoners, and migrants
- Reinfection risks without continued prevention

But never before has a virus gone from incurable to curable, then to eliminable, all within a single generation.

A Rare, Beautiful Word: Eradication

Smallpox is gone.

Polio is on the verge.

Hepatitis C could be next.

Because now we don't just have treatment.

We have a cure.

We have the will.

We have the science.

And for the first time, we have a future without HCV.

"I buried friends who had Hep C," said one advocate.

"I lived long enough to take the cure. That's not just science. That's salvation."

The Sofosbuvir Revolution

How One Pill Rewrote the Rules, Cured a Global Killer, and Sparked a New Era of Antiviral Medicine

For decades, Hepatitis C was a paradox.

It was a disease that millions carried but few talked about.

A silent infection of the liver that caused fatigue, fibrosis, and eventually failure, but often stayed invisible until the damage was irreversible.

Even worse, the treatment—interferon and ribavirin—was nearly as bad as the disease:

- Year-long injections
- Debilitating side effects: nausea, depression, anemia
- Low success rates, especially for certain genotypes

Patients called it "toxic hope."

And many chose to wait—for something better.

That "something" arrived in late 2013.

It was a small, white pill named Sofosbuvir.

And it didn't just improve Hepatitis C treatment.

It revolutionized it.

A Simple Pill. A Complex Journey.

Developed by a small biotech startup called Pharmasset, Sofosbuvir was based on a new idea:

Instead of boosting the immune system (like interferon), why not target the virus directly?

Hepatitis C needed a special enzyme—NS5B RNA polymerase—to make copies of itself.

Sofosbuvir mimicked one of the building blocks of the virus's RNA and stopped replication in its tracks.

It was:

- Oral (no injections)
- Fast-acting
- Highly potent
- And, crucially, genotype agnostic—it worked across most strains of Hep C

In 2011, Gilead Sciences acquired Pharmasset for $11 billion, betting everything on this molecule.

Two years later, Sovaldi (the brand name for Sofosbuvir) was approved.

The Results were Unbelievable

In clinical trials, Sofosbuvir-based regimens cured:

- 90–99% of patients
- In just 8 to 12 weeks
- With mild, almost negligible side effects

This was unheard of in virology.

There was no resistance, no long-term toxicity, and—most miraculous of all—no relapse.

"For the first time, we weren't suppressing a virus," said one physician.

"We were eliminating it."

What made it a Revolution

1. It was a cure.

   Not control. Not remission. A true virologic cure.

   After 12 weeks on Sofosbuvir and partner drugs, the virus was undetectable—and stayed gone.

2. It was simple.

   No injections. No refrigeration. A single pill, once a day.

   Even patients with cirrhosis, co-infections, or treatment failures could respond.

3. It was fast.

   No more year-long regimens.

   For many, 8 to 12 weeks was all it took.

4. It worked for almost everyone.

   Across all races, genders, geographies, and genotypes.

5. It opened the door to elimination.

   With the right political will and access, the virus itself could be wiped out from the global population.

From "Chronic and Progressive" to "Cured and Free"

Mohammed, a 48-year-old Egyptian taxi driver, had lived with Hep C for 15 years.

He had stopped working full-time, afraid of infecting his family.

Interferon had failed him.

Then he received 12 weeks of Sofosbuvir and daclatasvir through a government program.

"The virus is gone. My energy is back. I can live again," he said. "For me, this is a second life."

The Controversy: Hope Meets Cost

When Sovaldi launched in the U.S., the price stunned the world:

$1,000 per pill. $84,000 for a full course.

While the cost was defensible (years of R&D, high cure rates, saved lives), it triggered:

- Public outrage
- Debates over pharmaceutical pricing
- Calls for generic access in low-income countries

To Gilead's credit, they licensed Sofosbuvir to over 90 countries for low-cost generic production.

In India, the full course dropped to $300 or less.

This compromise made global access possible.

A Global Impact Measured in Lives

Since Sofosbuvir's approval:

- Millions have been cured globally
- Liver transplants dropped dramatically in Hep C patients

- Rates of liver cancer are falling in some regions
- National programs—from Egypt to Georgia—have scaled up mass treatment

It is one of the most effective and rapid public health victories in medical history.

The Legacy of Sofosbuvir

Sofosbuvir changed more than just Hepatitis C.

It:

- Set a new bar for drug efficacy
- Validated direct-acting antivirals
- Challenged pharma to rethink access and pricing models
- Inspired a new generation of short-course, curative therapies

It also gave people something rare in virology:

A finish line.

"I took one pill a day," said a retired nurse from Italy.

"Then I was cured. I still don't believe it. But the tests say it's true. I'm free."

A Pill that Opened the Door to Eradication

Sofosbuvir didn't just shrink a virus.

It sparked a movement to screen, treat, and eliminate Hepatitis C worldwide.

For the first time, doctors and health leaders aren't just talking about managing a disease.

They're talking about making it disappear—entirely.

And it all began with one pill, one enzyme, and one belief:

We can do better than treatment. We can offer a cure.

A Story of Redemption: "I was an Addict—Now I'm Virus-Free"

How the Cure for Hepatitis C became more than Medicine—It became a Chance to Start Over

Tyler doesn't remember the exact day he contracted Hepatitis C.

It could have been in a shared needle behind a gas station.

Or during the time he was sleeping in a friend's car, using anything to dull the pain of withdrawal.

Back then, the virus was the least of his worries.

"When you're addicted, you don't plan next week. You're not afraid of Hep C—you're afraid of getting sick before your next fix."

For years, he lived with the infection and didn't even know.

When he finally got tested at a rehab intake center, the result came back positive: HCV, genotype 1a.

The counselor told him, "It's treatable."

But Tyler laughed bitterly.

"I didn't believe in treatment. I didn't believe in myself. Not back then."

What he didn't know—what he couldn't know—is that he was standing at the edge of a story not about punishment, but about redemption.

Hepatitis C and Addiction: A Cruel Partnership

People who inject drugs account for the majority of new HCV infections in many countries.

It's not a moral failing. It's an epidemiological fact:

- Shared needles
- Unsterile conditions

- Lack of access to clean supplies or medical care
- And stigma that keeps people from testing

Many live for years unknowingly infected, only to discover it when liver damage sets in—or when they begin recovery.

Tyler's liver was already inflamed when he got his diagnosis.

He felt tired, anxious, and ashamed.

But he also felt something else—ready.

The First Step Toward Freedom wasn't just Detox. It was Trust

At a treatment center in Vancouver, Tyler met a nurse named Elena, who had helped hundreds through HCV treatment.

She told him about a new pill—Sofosbuvir, combined with Ledipasvir.

No injections. No horrible side effects.

Just one pill a day for 12 weeks.

"I asked if I had to be perfect—if one mistake would cancel the cure. She said, 'We don't treat the virus only if you're worthy. We treat it because you deserve to live.'"

That day, Tyler took his first dose.

Twelve Weeks of Recovery and Reckoning

Tyler stayed in rehab during treatment.

Each morning, he took his antiviral with coffee and joined a support group.

He cried on Day 11. Wanted to quit on Day 19. Laughed on Day 27.

By Week 6, his viral load was undetectable.

He couldn't believe it.

"After everything I'd done to my body… it was healing. That pill was doing what I never thought possible."

At Week 12: A New Life, Confirmed by Bloodwork

On the final day, Elena handed him his lab results.

"No detectable virus."

Tyler stared at the paper like it was a passport.

Not to another country—but to a future he thought he didn't deserve.

He was virus-free.

He was sober.

He was, for the first time in a decade, hopeful.

Why this Story Matters more than Ever

Because for too long, people with addiction were told:

- You're not compliant.
- You'll just get reinfected.
- You don't deserve the cure.

But science says otherwise:

- Studies show that people who inject drugs can be treated effectively
- That relapse does not negate the right to care
- That HCV treatment reduces transmission and is a form of prevention
- That healing the body often inspires healing the life

Tyler is one of millions.

But his story is also singular.

"I used to live like I didn't matter. That's what addiction does. Hep C treatment told me I do. That I'm worth curing."

Redemption is a Prescription—and a Choice

Today, Tyler:

- Sponsors others in recovery
- Speaks at harm reduction clinics
- Wears a bracelet that says, "Cured but not done."

His liver is healing.

His life is expanding.

His story is one of medicine—but also mercy.

“I was an addict. Now I’m virus-free,” he says.

“And if I deserve that chance, so does everyone else.”

Chapter 19

Breakthroughs in Rare Diseases

From Isolation to Innovation—When Medicine Remembered the Few

For most of medical history, if your disease affected only a few thousand people, you were invisible.

You were told:

"There's no treatment."

"There's no funding."

"There's no research interest."

These were the rare diseases—sometimes called orphan diseases.

Each one is too uncommon to command attention.

Collectively, too numerous to ignore.

Today, that story is changing.

A wave of breakthroughs has transformed rare disease care from resignation to resilience, from neglect to Nobel-worthy discovery.

This chapter is a tribute to those breakthroughs—and to the patients and families who refused to be forgotten.

Rare, but Not Rare

A disease is considered rare in the U.S. if it affects fewer than 200,000 people.

Globally, there are over 7,000 known rare diseases, and more are discovered each year.

Individually, they are uncommon.

Together, they affect over 300 million people worldwide—more than cancer or HIV.

Most rare diseases are:

- Genetic in origin
- Chronic, progressive, and life-threatening
- Diagnosed in childhood
- Historically, untreatable

But recent decades have brought hope—and headline-making science.

Turning Point: The Orphan Drug Act (1983)

It began with a bold legislative move.

The U.S. Orphan Drug Act incentivized companies to develop treatments for rare diseases by offering:

- Market exclusivity
- Tax credits
- Research grants
- Accelerated approval pathways

Since then:

- Over 1,000 orphan drugs have been approved
- Entire companies have been built around rare disease innovation
- Patients who once had no options are now seeing life-changing therapies

Landmark Breakthroughs that Changed the Landscape

Cystic Fibrosis: From Decline to Normalcy:

Once considered a childhood death sentence, cystic fibrosis (CF) is a genetic disease that thickens mucus, affecting the lungs and pancreas.

Then came CFTR modulators like Ivacaftor and Trikafta, which correct the defective protein itself.

"My son used to gasp for air," said one mother. "Now he plays soccer."

Children with CF now have a real shot at adulthood—and a full life.

Spinal Muscular Atrophy (SMA): From Wheelchairs to First Steps:

SMA was once the leading genetic cause of infant death.

Then came:

- Nusinersen (Spinraza) — an antisense therapy to increase survival motor neuron protein
- Zolgensma — a one-time gene therapy, replacing the faulty gene
- Risdiplam (Evrysdi) — an oral treatment improving motor function

"We were told our baby wouldn't live to walk," said one parent. "Now she's climbing stairs."

Pompe Disease: From Muscular Decline to Muscle Gain

Pompe is a rare enzyme deficiency that leads to progressive muscle weakness and heart failure.

In 2006, alglucosidase alfa (Myozyme) became the first enzyme replacement therapy for Pompe.

It was the story behind the movie Extraordinary Measures: a rare-disease breakthrough born from relentless parental advocacy and biotech innovation.

Paroxysmal Nocturnal Hemoglobinuria (PNH): From Fatal to Functional

PNH is a rare, life-threatening blood disorder.

Before 2007, patients often died young.

Eculizumab (Soliris), a monoclonal antibody, blocks the immune system's destruction of red blood cells.

It was one of the first complement inhibitors—paving the way for other rare blood disease therapies.

"I used to live with transfusions and fear," said one survivor. "Now I live normally."

Progeria: Unlocking Longevity, One Trial at a Time

Progeria causes accelerated aging in children, with most dying in their teens from cardiovascular disease.

In 2020, the FDA approved Zokinvy (lonafarnib), the first therapy to extend life in progeria patients based on decades of research, family advocacy, and sheer determination.

It added years of life and a sense of urgency and possibility to rare disease drug development.

The Rise of Precision Tools: Gene Therapy and Beyond

Many rare diseases are caused by a single gene mutation—making them ideal targets for gene therapy.

In the last decade, we've seen:

- Zolgensma for SMA
- Luxturna for inherited blindness
- Dozens of CRISPR trials for diseases like sickle cell, beta-thalassemia, and Fabry

The promise is bold:

Don't treat symptoms.

Fix the root cause.

Patients who Led the Way

Every one of these breakthroughs exists because of:

- Families who started foundations
- Patients who entered risky trials
- Parents who learned the science and lobbied pharma and government
- Doctors who refused to give up on one rare child

In rare disease, the patients aren't passive.

They're often the spark behind every success.

"We had to scream to be seen," said the mother of a child with Batten disease. "Now, people are listening."

What Breakthrough Really Means

In rare disease, a breakthrough isn't always a cure.

Sometimes it means:

- Delaying symptoms by years
- Turning a fatal illness into a manageable one
- Giving a parent one more birthday with their child
- Replacing despair with something to look forward to

The Road Ahead: From Rare to Recognized

Challenges remain:

- High costs—many therapies exceed six figures per year
- Long diagnostic odysseys
- Equity of access, especially in low-income settings
- Siloed research and slow trial enrollment

But we now have the tools. The will. The science.

And the stories.

"We used to be the forgotten families," said one parent.

"Now we're at the front of the line. And we're not just surviving. We're rewriting what's possible."

From Orphan Drugs to Life-Changing Therapies

How a Forgotten Category of Medicines became a Lifeline for the World's Rarest Patients

There was a time—not long ago—when having a rare disease was a life sentence not only of suffering, but of neglect.

The diagnosis came with six words more devastating than the disease itself:

"There is nothing we can do."

Drug companies called these conditions orphaned—too rare to profit from, too complex to solve, too small a market to matter.

But in the last four decades, something extraordinary happened.

A movement began.

A law was passed.

A market was created.

And science, fueled by purpose, rewrote the narrative.

This is the story of how orphan drugs became life-changing therapies—and how patients went from being left behind to leading the way.

The Problem: No Incentive, No Research, No Hope

In the 1970s, fewer than 10 drugs had ever been developed for rare diseases.

Companies had no financial reason to invest. The regulatory hurdles were high. The expected revenue was low.

Meanwhile:

- Children died of rare metabolic disorders
- Teens lost mobility to muscle-wasting diseases
- Adults battled symptoms with no name, no treatment, and no plan

There were over 7,000 rare diseases—but almost no treatments.

"We felt like we didn't matter," said one mother. "Medicine didn't know us. Pharma didn't see us."

The Turning Point: The Orphan Drug Act of 1983

In response to patient advocacy and moral urgency, the U.S. passed the Orphan Drug Act, offering:

- 7 years of market exclusivity
- Tax credits for clinical trials
- Waived FDA fees
- Grants for research

This law didn't just open a door—it flooded it with light.

Suddenly, what had been "too rare to matter" became a field of opportunity for innovation, compassion, and commercial viability.

A New Era Begins

By the 1990s and early 2000s:

- Startups emerged dedicated entirely to rare diseases
- Major pharmaceutical companies created orphan divisions
- The number of orphan drug designations soared into the thousands
- And most importantly, patients began getting answers

From Policy to People: Therapies that Changed Lives

Here are just a few examples of orphan drugs that transformed what it means to live with a rare disease:

Cystic Fibrosis

Ivacaftor, Lumacaftor, Tezacaftor, Elexacaftor (Trikafta)

These CFTR modulators didn't just ease symptoms—they improved lung function, reduced hospitalizations, and gave patients decades more life.

"I no longer plan my life around my illness," said one teen. "I plan my future."

Gaucher Disease

Imiglucerase (Cerezyme)

An enzyme replacement therapy that reversed organ damage, restored growth in children, and gave families back their time.

"My son was dying. Now he's in college."

Spinal Muscular Atrophy (SMA)

Nusinersen (Spinraza), Zolgensma, Risdiplam

These drugs transformed a universally fatal childhood disease into one where babies walk, talk, and live full lives.

"We were preparing to say goodbye. Now we're learning to ride bikes."

Hereditary Angioedema (HAE)

Lanadelumab (Takhzyro)

Frequent, life-threatening swellings—once unpredictable and disabling—are now nearly eliminated.

"I'm not afraid to go outside anymore."

The Economics of Empathy

Orphan drugs were once dismissed as unprofitable.

Today, they account for:

- Over 40% of new drug approvals
- Billions in revenue, which fuels further rare disease R&D
- The creation of entire companies—like Genzyme, BioMarin, and Sarepta—that exist solely to serve rare communities

This is not exploitation—it's the market doing something right:

Turning neglected conditions into investment-worthy missions.

Patients at the Center of Progress

Unlike in traditional drug development, in rare disease:

- Parents become researchers
- Patients help design trials
- Advocacy groups fund early science
- The line between patient and partner disappears

Their voices are not just heard.

They are often the reason the drug exists at all.

Challenges that Remain

- Cost: Many orphan drugs are expensive due to small markets and complex manufacturing

- Access: In low-income countries, these therapies remain out of reach
- Diagnosis: Many patients still wait 5–10 years for an accurate diagnosis
- Equity: Not all rare diseases attract equal attention—some remain in the shadows

But every orphan drug approved is proof of the possibility.

A signal to the world: If we can do this once, we can do it again.

A New Chapter: Gene Therapy, RNA Medicine, and CRISPR

The next wave of rare disease therapies may not just treat symptoms—they may correct mutations at the source.

And many of these breakthroughs began in the rare disease space, where the stakes are highest and the advocacy is strongest.

What was once an "orphan" field is now the future of medicine itself.

"They told us no one would care about our disease," said one advocate.

"We proved them wrong. Not just for us—but for everyone who comes next."

Children with SMA and the Miracle of Spinraza

How One Therapy Breathed Life into the Silent World of Spinal Muscular Atrophy

For decades, Spinal Muscular Atrophy (SMA) was one of the cruelest diagnoses a parent could hear.

It was a thief in slow motion—stealing movement, breath, and eventually life itself.

Children were born seemingly healthy, only to:

- Struggle to lift their heads
- Lose the ability to crawl or walk
- Fight for every breath as their muscles slowly failed

SMA Type 1, the most severe form, was a devastating sentence:

"Your child may not live past their second birthday."

There was no cure. No treatment. Only wheelchairs, ventilators, feeding tubes, and heartbreak.

Then came Spinraza and everything changed.

Understanding SMA: A Disease of a Missing Protein

SMA is caused by a mutation in the SMN1 gene, which encodes the survival motor neuron (SMN) protein, essential for muscle control.

Without it:

- Motor neurons in the spinal cord degenerate
- Muscles weaken and wither
- The body slowly disconnects from the brain's commands

But there's a twist:

Humans have a backup gene, called SMN2, which produces some SMN protein—just not enough to stop the disease.

Could we trick the SMN2 gene into making more of what's missing?

That question launched a revolution.

The Science behind the Breakthrough

In the early 2000s, scientists discovered how to manipulate RNA splicing—the cellular process that turns genetic instructions into functional protein.

A company called Ionis Pharmaceuticals, in partnership with Biogen, developed an antisense oligonucleotide—a tiny strand of synthetic RNA that would bind to SMN2 and tell it:

"Make the full-length, working SMN protein."

That molecule became nusinersen—later named Spinraza.

The Trials that Turned Tears into Hope

In clinical trials, infants with SMA Type 1 who received Spinraza:

- Moved muscles they had never moved before
- Sat, rolled, and in some cases walked
- Lived past their second birthday—and kept thriving
- Required less ventilatory support
- Hit developmental milestones no one thought possible

For the first time in history, children with SMA Type 1 were not declining. They were developing.

In 2016, Spinraza became the first-ever FDA-approved treatment for SMA.

The Miracle in the Clinic

Emma was six months old when her parents heard the words "SMA Type 1."

She couldn't hold her head up. Her breathing was shallow. Her pediatrician told them to prepare for palliative care.

But then she was offered Spinraza.

It was injected into her spinal fluid, a delicate procedure, done with trembling hope.

Weeks later, her mother noticed something incredible:

"She started moving her legs. Just a little at first. Then more. She smiled bigger. She held her bottle."

At nine months, she sat up.

At eighteen months, she took her first steps with a walker.

At three, she danced at her own birthday party.

"This isn't just a drug," her father said, holding back tears. "It's a second chance."

Beyond Type 1: Changing the Lives of Older Children

Spinraza is now approved for:

- All types of SMA (Types 1–4)
- Children and adults
- Patients on ventilators or with scoliosis

Even older kids who had lost significant strength showed improvements:

- More upper body movement
- Improved swallowing and breathing
- Greater endurance

For many, it slowed or halted progression, offering a stable future where only decline was expected.

Challenges and Courage

Spinraza must be:

- Delivered through lumbar puncture
- Given every 4 months indefinitely
- Priced at a premium—$750,000 in the first year, then $375,000 annually

But insurers, health systems, and governments around the world have stepped up because no cost compares to the value of a child's life.

The Dawn of a New SMA Era

Spinraza opened the floodgates for other therapies:

- Zolgensma: a one-time gene therapy
- Risdiplam (Evrysdi): a daily oral medication

But it was Spinraza that proved a new truth:

A genetic disease does not have to be a life sentence.

The Sound of a New Story

Today, in clinics around the world, you hear things that were once unimaginable:

- The giggle of a baby with Type 1
- The clatter of braces on a child running
- The rhythm of a wheelchair driven by strength, not surrender
- The words: "She's doing better than we ever dreamed."

"They told us to expect a funeral," said Emma's mother.

"Instead, we're planning preschool."

Chapter 20

COVID-19 Vaccines: A Race against Time

How Science, Speed, and Solidarity Delivered Hope to a World on its Knees

In early 2020, the world watched as an invisible enemy spread with terrifying speed.

Hospitals overflowed. Cities locked down. Borders closed.

Every cough sparked fear. Every update brought worse news.

The virus had a name: SARS-CoV-2.

The disease it caused—COVID-19—was not just a health crisis. It was a global trauma.

And as the death toll climbed into the millions, one desperate hope took center stage:

"We need a vaccine—and we need it now."

What followed was nothing short of a scientific sprint—an unprecedented, high-stakes race against time.

A Pandemic Demanding a Miracle

Traditional vaccine development takes 10–15 years.

But the world didn't have that kind of time.

The goal was unthinkable:

Develop, test, manufacture, and distribute safe, effective vaccines—within a year.

This would require:

- Global collaboration
- New technology
- Massive funding

- Regulatory flexibility
- And an unshakeable belief that science could catch up to the virus

The Front-Runners Step Forward

By March 2020, dozens of vaccine candidates were already in development.

By April, some were already in Phase I trials.

By the end of the year, a handful had made it across the finish line.

- **Pfizer–BioNTech (BNT162b2):** A German-American partnership using a new technology: mRNA.
- **Moderna (mRNA-1273):** A small biotech company that became a household name, also using mRNA.
- **Oxford–AstraZeneca (ChAdOx1):** A viral vector vaccine developed with academic speed and global reach.
- **Johnson & Johnson, Novavax, Sinovac, Covaxin:** Others soon followed with varying platforms: viral vectors, inactivated viruses, protein subunits.

What united them all was urgency—and hope.

The mRNA Revolution Arrives

Though mRNA vaccines had never been approved before, they'd been studied for years.

Instead of injecting the virus, mRNA vaccines delivered genetic instructions that taught the body to produce the spike protein, triggering an immune response.

Advantages:

- Faster to develop
- No need to grow virus cultures
- Highly adaptable to variants
- Stimulated both antibody and T-cell responses

Within months of sequencing the virus's genome, mRNA vaccines were in human arms.

And when the data came back, the world exhaled:

Pfizer–BioNTech: 95% effective.

Moderna: 94.1% effective.

These were stunning numbers, far better than anyone dared expect.

Regulatory Greenlights and a Global Mobilization

In December 2020:

- The UK approved Pfizer's vaccine—the first Western country to do so.
- The U.S. FDA followed within days.
- Other countries rapidly rolled out emergency authorizations.

Massive operations like Operation Warp Speed in the U.S. and COVAX globally began delivering millions of doses.

Cold chain storage, public messaging, safety surveillance—all scaled at unprecedented speed.

Clinics opened in parking lots, stadiums, and mosques.

Healthcare workers became vaccine ambassadors.

Elderly people wept at the first shots.

Volunteers lined up.

Billions watched and hoped.

Lives Saved, Curves Flattened

By mid-2021:

- COVID deaths plummeted in vaccinated populations
- Hospitalizations dropped sharply
- Millions returned to schools, weddings, and workplaces

Estimates suggest COVID vaccines saved tens of millions of lives worldwide in the first two years alone.

"It felt like a curtain lifting," said one nurse. "Suddenly, we weren't losing patients every shift."

Challenges Alongside Triumphs

The vaccine race succeeded—but not without struggle:

- Misinformation spread as fast as the virus
- Hesitancy and politicization divided communities

- Equity gaps emerged: rich nations had excess doses, while poorer ones waited
- Variants (Delta, Omicron) required rapid adaptation and booster strategies

But through it all, the vaccines held the line.

Even when they didn't prevent infection, they reduced severity, hospitalizations, and death.

Science's Finest Hour?

COVID-19 vaccines were:

- Developed in record time
- Based on decades of research
- Distributed to billions
- A proof of concept for rapid response to future pandemics

They turned biotechnology from theory into global salvation.

The Human Side of a Shot

For many, the vaccine was more than a jab.

It was:

- The chance to hug a grandchild
- To open a business
- To breathe without fear
- To mourn safely, live freely, and hope again

"I watched my dad die in the first wave," said one man.

"I watched my mom get her vaccine in the second. That's what victory looks like."

A Legacy of Preparedness

COVID-19 vaccines didn't end the pandemic alone.

But they proved:

- What is possible when the world acts together
- That crisis can catalyze transformation
- That science, when trusted and funded, can deliver miracles in real time

"This was a race," said Dr. Kizzmekia Corbett, one of the key developers of Moderna's vaccine.

"And we didn't just run. We flew."

How mRNA Vaccines Changed Science Forever

From a Pandemic Response to a Medical Revolution in Real Time

Before 2020, mRNA technology was the quiet promise of a few scientists, startups, and visionaries.

It had never been proven at scale.

It was dismissed as too unstable, too experimental, too niche.

Most people had never heard of it.

But when COVID-19 swept across the world, mRNA vaccines didn't just arrive—they arrived first. And they worked.

And with that, medicine changed forever.

What is mRNA and Why does it Matter

Messenger RNA (mRNA) is the blueprint your body uses to make proteins.

It carries instructions from DNA to the cell's protein-making machinery.

The idea behind mRNA vaccines is simple—but profound:

Instead of injecting a protein (like traditional vaccines), give the body the instructions to make it itself just enough to trigger an immune response, and then disappear.

That means:

- No virus needs to be grown
- No risk of infection
- Rapid design and adaptability to new variants
- Clear, targeted immune stimulation

The concept had been explored for decades. But no mRNA vaccine had ever been approved—until Pfizer–BioNTech and Moderna shattered that barrier.

Why the COVID Breakthrough was Historic

When the SARS-CoV-2 genome was published in January 2020:

- Moderna designed their vaccine candidate within 48 hours

- Pfizer–BioNTech followed with a similar approach
- Both entered human trials within months

And when the results came in:

- Pfizer–BioNTech showed 95% efficacy
- Moderna showed 94.1% efficacy

These were some of the most effective vaccines ever developed—achieved in less than a year.

For the first time, mRNA proved itself on a global stage:

- Scalable
- Safe
- Powerful
- Flexible

A Platform, Not just a Product

What made mRNA revolutionary wasn't just COVID.

It was that mRNA became a platform—a modular, programmable foundation that could be adapted to fight any protein-based disease.

It shifted vaccine development from biology to bioengineering.

With mRNA, scientists can now:

- Design a vaccine candidate in days
- Manufacture it without cell cultures or eggs
- Update it quickly to match emerging variants

This is speed medicine.

Digital biology.

Vaccines on demand.

What mRNA Unlocks Next

The success of mRNA in COVID accelerated efforts across dozens of fields:

- **Flu Vaccines:** mRNA can adapt seasonally to dominant strains, potentially offering better and longer-lasting protection than traditional flu shots.

- **RSV and CMV:** Early trials show strong immune responses against these common but dangerous viruses, especially in infants and the elderly.
- **HIV:** With mRNA, scientists can test dozens of envelope protein configurations rapidly—a task that once took years.
- **Malaria, TB, Zika**: Diseases that once eluded traditional vaccine approaches are now being reimagined through mRNA science.
- **Cancer Vaccines:** This is perhaps the most breathtaking promise. Personalized cancer vaccines using mRNA to train the immune system to attack tumor-specific neoantigens.

Companies like Moderna and BioNTech are already in advanced trials:

- Patients 'tumors are sequenced
- Unique cancer markers identified
- Custom mRNA vaccines are generated—tailored to each patient's tumor

Early results in melanoma and pancreatic cancer are promising.

"It's like sending your immune system to sniper school," said one oncologist.

Rare Diseases and Protein Deficiencies

For disorders caused by missing or faulty proteins (like certain enzyme deficiencies), mRNA offers the potential to:

- Deliver temporary, regulated protein production
- Provide episodic or continuous treatment without gene editing

This could dramatically alter the landscape for inherited metabolic diseases.

Beyond Vaccines: A New Chapter in Therapeutics

mRNA isn't just for preventing disease—it's being developed for treating disease:

- Cardiovascular medicine: mRNA that prompts growth of blood vessels
- Autoimmune regulation: mRNA to induce immune tolerance in Type 1 diabetes and multiple sclerosis
- Antibody replacement: mRNA that codes for monoclonal antibodies inside the body

It's not just a molecule.

It's a medical language, one we're just beginning to speak fluently.

The Legacy of the Pandemic's most Powerful Tool

The COVID-19 crisis showed what mRNA can do:

- Save millions of lives
- Be scaled to billions of doses
- Be trusted, tweaked, and tuned in real time

It collapsed timelines.

It dismantled disbelief.

And it planted the seeds for a future of programmable medicine.

"We always believed in mRNA," said Dr. Katalin Karikó, a pioneer of the field.

"Now the world believes too."

From Experimental to Essential

What was once a speculative field is now:

- A cornerstone of pandemic preparedness
- A frontline tool in immunology
- A hopeful horizon for diseases we once couldn't touch

mRNA vaccines didn't just end a pandemic.

They began a new era of medicine, one built not just on hope, but on code.

A Global Story of Hope and Resilience

How a World Brought to its Knees Chose Science, Unity, and the Unseen Strength of the Human Spirit

In early 2020, the world stood still.

Airports emptied.

Schools closed.

Silence settled over cities that once pulsed with life.

The virus—new, nameless at first—spread fast. Too fast.

First, it was Wuhan. Then Milan. Then New York. Then everywhere.

It didn't care about borders, wealth, religion, or rank.

We lost lives. We lost livelihoods. We lost our sense of control.

But we did not lose one thing:

The will to survive. And the courage to keep going.

The Shock was Global—But so was the Response

In every country, something remarkable happened.

Doctors, nurses, and first responders showed up in overwhelmed hospitals with no roadmaps, just resolve.

Scientists stayed up through the night, sharing open-source genome data across continents, chasing down treatments, testing ideas, and daring to believe a vaccine could come faster than ever before.

Entire industries retooled:

- Perfume factories made sanitizer
- Auto plants made ventilators
- AI labs screened antiviral compounds
- Governments funded research not in years, but in days

It was a war effort. But our weapons were data, microscopes, and human empathy.

Acts of Kindness in the Darkest Times

In slums, soup kitchens popped up overnight.

Teenagers delivered groceries to the elderly.

Musicians played concerts from balconies.

People lit candles in their windows for those they had lost—and for those still fighting.

Some wore full PPE. Others wore fear on their faces.

But across every culture and every language, there was one unspoken promise:

"We will not let go of each other."

Science Delivered—But Humanity Carried Us

The vaccines arrived faster than anyone thought possible.

But they weren't enough on their own.

What truly changed the course was:

- The trust of communities
- The bravery of trial volunteers
- The solidarity of nations sharing doses
- The grace of people waiting for their turn

From Brazil to Bangladesh, New Zealand to Nigeria, people stood in lines, not for bread, but for protection.

"This isn't just a shot in the arm," onc health worker said. "It's a signal of hope."

Children Learned at Kitchen Tables. Elders Learned to Video Call.

We adapted in ways we never imagined:

- Weddings on Zoom
- Birthdays celebrated through glass
- Funerals streamed from far away

And yet, even distanced, we were never truly alone.

Global Health was Redefined

COVAX sent millions of doses to countries that couldn't afford them.

Cold chains reached mountain villages.

Mobile clinics vaccinated nomadic communities.

Public health workers became heroes without capes, armed with ice boxes and hope.

We realized health is not local. It is universal.

A virus anywhere is a threat everywhere.

But so too is a cure.

What Resilience Really Means

Resilience wasn't loud.

It was:

- A single mom turning her living room into a classroom
- A doctor whispering encouragement through a face shield
- A factory worker doubling shifts to keep the world supplied
- A teenager helping grandparents book their vaccine slots
- A nurse who held a stranger's hand so they wouldn't die alone

Resilience was choosing compassion over convenience, and science over cynicism.

Not Everything was Healed—but Everything was Touched

We lost much. And we carry those scars.

But we gained too:

- A renewed faith in science
- A clearer view of inequity
- A deeper respect for caregivers
- A recognition that progress is possible—when we do it together

The World Bent, but did not Break

The virus came for our lungs.

But it could never take our breath entirely.

We filled that breath with action. With learning. With love.

And when the wave passed, we emerged not the same—but stronger in ways we never knew we needed to be.

"This was not the end of the world," said one survivor.

"It was the end of pretending we aren't connected."

Hope is not just a Feeling. It's a Practice

And in this pandemic, it looked like:

- A scientist in Germany who coded a vaccine at her kitchen table
- A truck driver who delivered it overnight across frozen roads
- A grandmother who finally hugged her grandchild
- A world that, for once, looked in the same direction—and acted

This was a tragedy.

But it was also a triumph:

Of medicine.

Of movement.

Of each other.

This was the global story of us.

Not just how we survived the virus.

But how we chose life and chose each other over and over again.

PART VI: THE FUTURE PHARMA IN BUILDING (2020s AND BEYOND)

Part VI

The Future Pharma is Building (2020s and Beyond)

From Reactive to Predictive, From One-Size to Personal, From Molecule to Mission

After decades of discovery, formulation, and prescription, the pharmaceutical industry has reached a turning point.

The past was about blockbusters and mass production.

The future is about precision, prevention, personalization, and purpose.

No longer is pharma just about treating disease.

Now, it's about anticipating it, targeting it, preventing it, and transforming the very definition of health.

The 2020s have become a launchpad, not just for new drugs, but for an entirely new era of medicine.

1. **Precision Medicine: Treatment Tailored to You**

   Gone are the days when everyone with the same diagnosis received the same pill.

   Today's pharma is building therapies based on:

   - Your genes
   - Your immune profile
   - Your tumor's exact mutations
   - Even your microbiome, digital behavior, and real-time health data

   Companion diagnostics, liquid biopsies, and AI-powered analytics are turning each patient into a roadmap.

   And with that roadmap, pharma is delivering:

   - Custom cancer vaccines
   - Targeted biologics
   - Pharmacogenomic-informed drug choices

- And therapies designed not just for populations—but for individual people

2. **mRNA, Gene Therapy, and Beyond: Programming the Body to Heal**

   COVID didn't just fast-track mRNA. It validated it.

   Now, that same platform is expanding into:

   - Pan-coronavirus vaccines
   - Personalized cancer therapies
   - Flu, HIV, and malaria prevention
   - Rare diseases once thought untreatable

   Meanwhile, gene therapy and gene editing (like CRISPR) are rewriting the rules—literally.

   In the coming decade, pharma aims to:

   - Cure monogenic diseases with one-time gene infusions
   - Edit faulty genes in vivo
   - Silence disease-driving mutations with RNA tools

   This is pharma not as pharmacy, but as bioengineering lab.

3. **Digital Therapeutics and AI-Driven Drug Discovery**

   The future of pharma isn't just chemical; it's computational.

   AI now helps:

   - Predict drug interactions
   - Simulate clinical trials
   - Analyze protein folding (thanks to DeepMind's AlphaFold)
   - Discover novel drug candidates in weeks, not years

   And digital therapeutics—prescription apps that manage disease—are being developed alongside pills.

   "In the future, your medication may come with an algorithm," says one biotech CEO.

4. **Pharma's Expanding Purpose: From Sick Care to Healthspan**

   Tomorrow's pharma will not just manage disease—it will optimize well-being, resilience, and longevity.

   Expect advances in:

   - Neuroregeneration for Alzheimer's and Parkinson's
   - Senolytics to slow aging and extend vitality

- Mental health innovations using psychedelics and neurotech
- Preventive therapeutics based on real-time biomarker surveillance

We're shifting from episodic care to continuous, proactive support—where pharma keeps you well, not just treats you when you're sick.

5. **Global Equity and Access: Pharma with a Conscience**

The pandemic exposed brutal inequities. But it also inspired unprecedented collaboration:

- Pooled procurement through COVAX
- Patent sharing platforms
- mRNA manufacturing in Africa
- Tiered pricing models for gene therapies

The future pharma is building will be judged not only by its innovation, but by its inclusion.

"A medicine that doesn't reach the people who need it is not a solution. It's an academic exercise."

Pharma's greatest opportunity lies not in the next molecule, but in the next billion lives impacted.

6. **Regenerative and Cell-Based Therapies**

We're entering an age where pharma no longer just fights disease; it rebuilds tissue.

This includes:

- CAR-T therapies that engineer your immune system to kill cancer
- Stem cell treatments for spinal injuries and heart disease
- Bio-printed tissues and organs under development

The old pharma offered pills.

The new pharma may offer cells, circuits, and even synthetic organs.

7. **Sustainability and Responsibility in Innovation**

Tomorrow's pharma must also be:

- Green: sustainable supply chains and biodegradable excipients
- Transparent: real-world evidence and open-label trials
- Ethical: AI that respects privacy, algorithms free from bias, clinical trials that reflect global diversity

The industry is being reshaped by public trust, environmental stewardship, and shared value.

What does it All Mean?

The pharma of tomorrow will be:

- Faster than ever before
- Smarter, powered by AI, genetics, and patient data
- Kinder, driven by patient experience and purpose
- More equitable, reaching more lives with more empathy

It will heal not just cells, but systems.

It will connect not just treatments, but people, across continents and conditions, through shared hope.

It will be a force not just for longevity, but for human flourishing.

"We used to ask: Can we make a drug for this?"

"Now we ask: Can we make a better world through this?"

The Future is not Fiction—It's already here

You'll see it in a child cured of SMA.

In a cancer patient receiving a vaccine made just for them.

In a young adult who never gets HIV because of a long-acting preventive injection.

In a family in a rural village, receiving gene therapy through a drone-drop delivery.

That's not science fiction.

That's the future pharma is building and it's happening now.

From molecules to miracles.

From products to promises.

From treatment to transformation.

Pharma is not just catching up to the future; it's shaping it.

Chapter 21

Gene Therapy and CRISPR

How we Moved from Treating Symptoms to Editing the Code of Life

For centuries, medicine treated what it could see:

Symptoms. Signs. Scars. Syndromes.

But now, we've entered a radical new era—one where we go beneath the surface, into the blueprint itself, to rewrite what's broken, silence what's harmful, or restore what was missing.

That blueprint is DNA, and the tools we now have—gene therapy and CRISPR—are giving us the power not just to treat disease, but to correct it at its root.

What is Gene Therapy?

Gene therapy is the delivery of new or modified genetic material into a person's cells to replace, fix, or supplement a faulty gene.

It's a molecular message that says:

"Here's the part you were missing. Now, make it work."

Gene therapy can be:

- Gene addition: inserting a healthy copy of a gene
- Gene editing: repairing a mutation within the gene
- Gene silencing: shutting down a toxic gene expression
- Gene replacement: removing a faulty gene and swapping in a functional version

It's typically delivered using viral vectors—modified viruses that act like couriers, carrying the correct genetic instructions into a patient's cells.

From Idea to Impact: The Breakthroughs Begin

For years, gene therapy was a dream delayed—hindered by setbacks, immune reactions, and technical limits.

But then came Zolgensma, approved in 2019 for Spinal Muscular Atrophy (SMA).

A single IV infusion that replaces the missing SMN1 gene—and has changed the lives of children who once would not have lived past two.

"One shot, and my child began to breathe on her own," said one mother. "Gene therapy didn't just change the disease; it changed our destiny."

Other gene therapy breakthroughs followed:

- Luxturna: restoring vision in inherited retinal disease
- Valoctocogene roxaparvovec: treating severe hemophilia A
- LentiGlobin: potentially curing sickle cell disease and beta thalassemia

In every case, patients didn't just improve—they were liberated from diseases that once defined their lives.

Then Came CRISPR: The Scalpel of the Genome

If gene therapy is like rewriting an entire chapter, CRISPR is a precise editor that finds a single typo in the genetic text—and corrects it.

Discovered in 2012 by scientists including Jennifer Doudna and Emmanuelle Charpentier, CRISPR-Cas9 gives researchers the ability to:

- Cut DNA at a specific location
- Remove, replace, or disable a gene with astonishing accuracy
- Do it quickly, cheaply, and across almost any organism

It was called:

"The most important discovery in biology since the double helix."

In 2020, Doudna and Charpentier won the Nobel Prize and CRISPR moved from the lab into the clinic.

From Mutation to Cure: CRISPR in Real Lives

In 2023, two patients with sickle cell disease were declared functionally cured after receiving CRISPR-based therapy that reactivated fetal hemoglobin.

One of them, Victoria Gray, had lived with agonizing pain crises her whole life.

After the treatment, she said: “It's like I've been reborn.”

Now, trials are underway for:

- Beta-thalassemia
- Hereditary angioedema
- Rare eye diseases
- Cancers
- HIV and inherited heart conditions

The future? Even in vivo CRISPR—editing genes directly inside the body, in real time.

Ethical Frontiers: What can we do, and what should we?

With great power comes great responsibility.

Gene editing raises difficult questions:

- What's the line between therapy and enhancement?
- Should edits be passed to future generations?
- Who gets access and who decides?

After China's controversial use of CRISPR to create genetically modified babies in 2018, the world drew a red line:

Germline editing—changes passed to future generations—must be approached with extreme caution.

But in the realm of somatic gene editing—treating a person without affecting their offspring—a global consensus is forming:

Do it, but do it ethically.

Pharma's Role in the Genetic Revolution

Pharmaceutical companies are now:

- Partnering with biotech pioneers in CRISPR and gene therapy
- Investing in delivery systems, like lipid nanoparticles and viral vectors
- Developing platforms for rapid gene-based drug development
- Building out manufacturing for personalized therapies that didn't exist a decade ago

This isn't just innovation; it's reinvention.

Pharma is no longer just discovering drugs.

It's designing genetic solutions, one patient at a time.

What this Means for Patients

Imagine a world where:

- A child born with a rare disease receives a single curative dose before symptoms appear
- A cancer patient's tumor is profiled and edited in real time
- A person with lifelong pain from sickle cell disease receives a single therapy—and is pain-free forever

That world is not decades away.

It's already beginning.

Gene by Gene, Future by Future

Gene therapy and CRISPR represent more than technology.

They are:

- A reversal of fate for millions
- A testament to human ingenuity
- A step toward a future where illness is not destiny

They're not just editing DNA.

They're rewriting what medicine can be.

"We don't just treat disease anymore," said one gene therapy pioneer.

"We go to the source. And fix it."

From Inherited Tragedy to Genetic Triumph

When a Family's Pain Meets the Future—and Everything Changes

For generations, inherited diseases carried an unspoken legacy:

Not just genes, but grief.

Not just DNA, but despair.

Hemophilia.

Sickle cell disease.

Duchenne muscular dystrophy.

Tay-Sachs.

Spinal muscular atrophy.

These weren't just medical conditions.

They were family stories of funerals too early, wheelchairs too soon, hopes too often buried with a birth certificate.

But in this century, quietly, bravely, brilliantly—that story has begun to change.

The Curse in the Bloodline

Marcus was born with sickle cell disease, just like his older brother, who died at 11.

As a child, Marcus screamed through pain crises so severe that his bones felt like they were cracking from the inside out.

His mother called them "silent storms"—nothing she could do but hold him through it.

"This is our life," the doctor told her. "Our inheritance."

They lived around hospitals. School was interrupted. Dreams were scaled back.

And then one day, she heard two words she never thought she'd hear in the same sentence:

"Gene therapy" and "cure."

The Science Meets the Story

Marcus entered a trial. He received his own stem cells, genetically edited using CRISPR to switch on fetal hemoglobin, undoing the mutation, silencing the pain.

Within months:

- No more ER visits
- No more opioids
- No more pain crises

"I had spent 20 years carrying this disease," Marcus said. "Now, I carry possibility."

He wasn't just a patient. He became the first generation to break the chain.

Duchenne: When Muscles Fade—Until they don't

Luca was diagnosed with Duchenne muscular dystrophy at age four.

His parents were told he'd be in a wheelchair by 10, and likely wouldn't live past 20.

But gene therapy was being trialed, delivering a micro-dystrophin gene to help restore muscle function.

He received an infusion as part of a clinical trial.

Months later, his parents watched, with tears and disbelief, as he climbed a staircase on his own.

"I used to pray he'd keep walking. Now I pray he'll run."

Inherited Tragedies Rewritten

- Hemophilia, once defined by internal bleeding and short lifespans, now has gene therapy candidates offering sustained clotting factor correction.
- Retinal dystrophies, which once led to blindness in childhood, are now being reversed with Luxturna.
- Beta thalassemia, once requiring lifelong transfusions, is being functionally cured with gene editing.

These aren't one-offs.

They're signs of a paradigm shift—from managing symptoms to correcting the biological script.

What Families say Now

"We lost three children before we even had a diagnosis. Now our son has a treatment."

"I carried guilt for passing this gene on. Now I carry gratitude."

"This isn't just science. It's healing generations."

Where once families feared testing—afraid of what might be revealed—they now seek answers.

Because answers may now come with options.

From Inheritance to Empowerment

Genetic counseling, carrier screening, and early intervention are being paired with therapies that:

- Replace missing enzymes
- Silence harmful mutations
- Stimulate beneficial pathways
- Or edit the genome directly

What was once a locked fate is now an open door.

And Still, the Work Continues

This triumph is not yet universal:

- Many rare conditions still lack effective treatments
- Gene therapies remain expensive and complex
- Global access is far from equal
- The emotional toll of living in limbo—between mutation and medicine—is very real

But the path forward is clearer than ever.

And it leads to something once unimaginable:

A world where a diagnosis doesn't begin the grieving—it begins the healing.

Legacy Rewritten

For so long, inherited disease meant passing down loss.

Now, families like Marcus's, Luca's, and thousands more are passing down something new:

- Clinical trial enrollment letters

- Normal lab results
- Graduation gowns
- Wedding photos
- And the most precious heirloom of all—hope

"We always thought this was the end of the story," one mother said.

"Turns out, it was the beginning of a miracle."

Stories of Lives Changed by Luxturna and Zolgensma

When One Infusion Meant Sight. When One Shot Meant Survival.

For most of modern medicine, doctors could treat symptoms.

Alleviate pain. Slow decline.

But reverse a genetic condition?

Restore what was lost before life had barely begun?

That belonged to science fiction.

And then, in the span of just a few years, two stories emerged—one of sight, one of strength—each made possible by gene therapy.

This is the story of Luxturna and Zolgensma—and the lives they transformed.

Luxturna: The Gift of Sight, Delivered by a Virus

Leber congenital amaurosis (LCA) is a rare inherited retinal disorder.

Children with the RPE65 mutation are often born with poor vision and face complete blindness in early childhood.

No glasses could help.

No surgery could fix it.

But in 2017, Luxturna (voretigene neparvovec) became the first FDA-approved gene therapy for an inherited disease.

It delivers a working copy of the RPE65 gene directly into retinal cells via a viral vector, enabling the eye to process light properly.

Michael: From Shadows to Fireflies

At age 8, Michael had never seen stars. He learned to walk by feeling the edge of walls. He played inside during recess because sunlight made it harder to see.

His parents knew that soon, he'd go completely blind.

Then came Luxturna.

A team of surgeons delicately injected the gene therapy behind his retina, one eye at a time.

Three weeks later, something astonishing happened.

"He ran outside and said, 'There are bugs with lights!'" his mother recalled.

"He had never seen a firefly."

Today, Michael:

- Reads large print
- Rides a bike with assistance
- Navigates with confidence
- And yes—stargazes

"Luxturna didn't just give him vision," his father said.

"It gave him independence. And a future."

Other Families Echo the Miracle

- Emily, 13, now reads printed books after Luxturna.
- Devon, 9, sees faces clearly for the first time.
- Sofia, 16, avoids a school for the blind and stays in mainstream education.

Where once there was darkness—now, there's light.

Not perfectly. Not fully. But enough to change a life forever.

Zolgensma: Strength Delivered in a Syringe

Spinal Muscular Atrophy (SMA Type 1) is a genetic disease that destroys motor neurons in infants— stealing the ability to move, breathe, or swallow.

Without treatment, most children don't live past age two.

Then came Zolgensma, approved in 2019.

A one-time IV infusion that delivers a functional SMN1 gene, replacing the faulty one.

It doesn't slow the disease.

It changes its trajectory entirely.

Avery: From Lifeless to Laughing

Avery was 7 months old when her pediatrician noticed she was no longer kicking her legs. She wasn't rolling. Her arms flopped like they didn't belong to her.

The diagnosis was SMA Type 1. Her parents were told to prepare for palliative care.

Then they heard about Zolgensma, a gene therapy available only through an expanded access program.

They applied. They waited. They cried.

And then, she was approved.

Within months of receiving the infusion:

- Avery could hold her head up
- Then sit
- Then crawl
- And eventually, stand with help

"We watched her come back to us," her mother said.

"We were told to prepare for goodbye. Now we're planning her third birthday."

One Shot, One Future

Zolgensma has now been given to thousands of children worldwide:

- Leo, once fed through a tube, now eats birthday cake on his own
- Noah, once immobile, now plays with siblings
- Alina, once dependent on oxygen, now sings songs without stopping to gasp

It's not a cure, but for many, it's a lifeline and a launching pad.

From Despair to Possibility

Luxturna and Zolgensma are more than just medical breakthroughs.

They are emblems of a new era, one where genetic fate is no longer final.

They've proven:

- That gene therapy can work

- That early intervention changes everything
- That children born with devastating diagnoses can live, learn, laugh, and lead

What Families say Now

"We went from planning hospice to planning preschool."

"She was fading. Now she's thriving."

"We thought he'd never see our faces. Now he looks us in the eyes and smiles."

"We thought this gene meant the end. Now it's just a chapter."

The Legacy of the Firsts

Luxturna and Zolgensma were firsts and they opened the floodgates.

Because of them:

- Regulators trust gene therapy more
- Investors fund more rare disease programs
- Doctors test earlier
- Families believe again

They didn't just treat conditions. They transformed conviction—what we think is possible in medicine, and in life.

"One vial gave my child strength."

"One drop gave my child sight."

"One shot gave us hope."

Chapter 22

Immunotherapy and the Power of the Body to Heal Itself

When Medicine Stepped Aside, and the Immune System Stepped In

For much of medical history, the immune system was a mysterious force.

Doctors respected it, but didn't dare touch it.

It could save your life—or destroy it in a blaze of inflammation.

In cancer, it often failed entirely—blind to the danger growing inside.

But then came a radical idea:

What if we could train the immune system to see what it's missing?

What if healing didn't have to come from outside, but from within?

That idea became immunotherapy.

And it didn't just change outcomes. It changed the story—from helpless decline to immune-driven defiance.

The Body's Forgotten Superpower

Your immune system is a marvel:

- It patrols every corner of your body
- It remembers past infections
- It adapts, evolves, and escalates when necessary
- It can kill viruses, reject organs, and trigger fevers with pinpoint precision

But for all its brilliance, it sometimes fails:

- Cancer cells disguise themselves as "self"
- Chronic diseases silence immune alarm bells
- Autoimmune conditions send it into overdrive

Immunotherapy doesn't replace the immune system.

It re-educates it.

Checkpoint Inhibitors: Taking the Brakes Off

In many cancers, immune cells are present, but paralyzed.

The cancer expresses proteins like PD-L1, which bind to PD-1 receptors on T-cells and deliver a chilling message:

"Don't attack. I'm one of you."

Drugs like nivolumab (Opdivo) and pembrolizumab (Keytruda) block this signal, freeing the T-cells to do their job.

This approach, called checkpoint inhibition, has revolutionized treatment for:

- Melanoma
- Lung cancer
- Bladder cancer
- Kidney cancer
- Head and neck cancers
- Hodgkin's lymphoma
- And many others

Emily: From Terminal to Triumph

Emily, 34, was diagnosed with stage IV non-small cell lung cancer. She didn't smoke. She had two young daughters. The prognosis was grim.

Then she was tested for PD-L1. The tumor expressed it at 90% meaning she was a perfect candidate for immunotherapy.

She received Keytruda, one infusion every 3 weeks.

Six months later, her scans showed a 70% reduction in tumor burden.

By year one, her cancer was no longer visible on imaging.

"I didn't just survive," she says. "I reclaimed my life."

CAR-T Cell Therapy: Engineering a Living Cure

Another form of immunotherapy takes healing even further.

In CAR-T therapy, a patient's T-cells are:

1. Removed from the body
2. Genetically modified to recognize cancer
3. Multiplied in the lab
4. And infused back, ready to hunt and kill

These supercharged cells have:

- Cured children with leukemia
- Put multiple myeloma into remission
- Given new options to patients with relapsed lymphomas

They don't just treat the disease; they fight for the patient long after infusion.

The Ripple Effects: Beyond Cancer

Immunotherapy is now being explored in:

- Autoimmune diseases: resetting immune balance in MS, lupus, Type 1 diabetes
- Allergies: retraining the immune system to stop overreacting
- Infectious diseases: enhancing immunity to HIV, TB, and chronic viral infections
- Neurodegenerative diseases: clearing amyloid plaques in Alzheimer's via antibody therapies

We're not just treating illness.

We're learning to restore immune harmony.

The Body: Teacher, Student, Warrior, Healer

Immunotherapy proved what once seemed impossible:

- That melanoma could disappear
- That leukemia could be cured in a single infusion
- That even advanced-stage cancers could be brought to a halt by the body itself

But more than that, it reshaped our faith, not just in medicine, but in the brilliance of the body.

Challenges and Caution

Immunotherapy isn't without risks:

- It can trigger autoimmune reactions
- It doesn't work for every patient
- It can cause immune storms that must be carefully managed

But in the right hands, and for the right patient, it can turn a grim prognosis into a second chance.

The Future: Smarter, Safer, Deeper Immune Design

Scientists are now working to:

- Combine immunotherapy with radiation, chemo, or targeted drugs
- Predict who will respond using biomarkers and AI
- Create off-the-shelf CAR-T cells
- Develop vaccines for cancer—training the immune system before disease even starts

This is not medicine of the past.

It is the medicine of the next century, arriving early.

"We didn't add a chemical to kill the tumor," said one oncologist.

"We reminded the body what the enemy looked like—and it took care of the rest."

A Final Thought: Healing, from Within

Immunotherapy teaches us that inside every patient is not just weakness, but wisdom.

A power forged over millions of years of evolution.

A system that, when guided, doesn't just fight back; it fights with purpose.

In this revolution, drugs don't heal the body.

The body heals itself.

And there is nothing more miraculous than that.

The Evolution of Checkpoint Inhibitors

How Unlocking One Immune Pathway Turned Despair into Defiance—and gave Cancer Patients a Fighting Chance

For decades, cancer treatment followed a brutal logic:

Kill the cancer before it kills the patient.

Chemotherapy. Radiation. Surgery.

Blunt tools for a clever enemy.

Then came an idea that changed everything.

What if we didn't kill the cancer directly?

What if we simply removed the brakes from the body's own immune system—and let it do what it was always meant to?

This was the dawn of checkpoint inhibition—a field that would not only revolutionize cancer care but alter how we understand healing itself.

Immune Checkpoints: The Body's Brakes

The immune system is powerful, but it's not reckless.

It's built with checkpoints—molecular signals that slow down or shut off immune responses when they're no longer needed.

These checkpoints help prevent:

- Autoimmunity
- Excessive inflammation
- Friendly fire on healthy cells

But cancer, in its cunning, hijacks these brakes.

One of the most notorious pathways?

PD-1/PD-L1

- PD-1: a receptor on T-cells
- PD-L1: a ligand on cancer cells

When they bind, the T-cell gets a message:

"Don't attack. I'm normal."

This allowed tumors to hide in plain sight even from the immune system designed to destroy them.

Early Clues and Long Resistance

In the 1990s, researchers like Tasaki Honjo in Japan discovered PD-1, and Jim Allison in the U.S. studied CTLA-4, another immune brake.

They suspected that blocking these pathways could reinvigorate T-cells—but the idea was met with skepticism.

Why?

- Cancer wasn't seen as an immune disease
- Prior immune therapies had failed or caused toxicity
- Drug developers were hesitant to bet on immunology

But Allison's work on CTLA-4 eventually led to the development of ipilimumab—the first checkpoint inhibitor to enter clinical trials.

The First Breakthrough: Ipilimumab and Melanoma

In 2011, the FDA approved ipilimumab (Yervoy) for metastatic melanoma.

It wasn't perfect—but it doubled survival in a disease where 6-month death was standard.

Patients who had run out of options were living for years. Some were even functionally cured.

"I expected to die in 2010," said one survivor. "I'm celebrating my daughter's graduation."

It wasn't just a drug.

It was a paradigm shift.

PD-1 Inhibitors: The True Revolution Begins

The real explosion came with drugs that blocked PD-1 or PD-L1:

- Nivolumab (Opdivo) – approved in 2014
- Pembrolizumab (Keytruda) – approved in 2014
- Atezolizumab (Tecentriq) – approved in 2016

These checkpoint inhibitors freed T-cells from paralysis, allowing them to recognize and destroy cancer cells.

And they worked across multiple cancers:

- Non-small cell lung cancer
- Kidney cancer
- Head and neck cancers
- Bladder cancer
- Triple-negative breast cancer
- Hodgkin lymphoma
- Microsatellite instability–high (MSI-H) tumors — regardless of origin (a first in oncology)

The Famous Case: Jimmy Carter

In 2015, former U.S. President Jimmy Carter was diagnosed with metastatic melanoma that had spread to his brain.

He received Keytruda and went into remission.

"I'm cancer-free," he announced to the world.

A moment that transformed public trust in immunotherapy.

Combination Therapies and New Frontiers

Researchers found that combining PD-1 inhibitors with CTLA-4 inhibitors (like Opdivo + Yervoy) produced even stronger responses, especially in:

- Melanoma
- Renal cell carcinoma
- Lung cancer

Other combinations with chemo, targeted therapy, and radiation are extending benefits to more patients.

Today, checkpoint inhibitors are being tested in over 3,000 active trials, including in:

- Colorectal cancer
- Pancreatic cancer
- Glioblastoma
- Ovarian cancer

What Patients Say: From End Stage to Encore

"I went from hospice to hiking."

"My doctor said we were out of options. Then he said: 'Let's try one more thing.'"

"I saw my daughter get married. That was never supposed to happen."

For many, these drugs didn't just shrink tumors; they restored dignity, time, and possibility.

Limitations and Lessons

Checkpoint inhibitors aren't for everyone. Challenges remain:

- Only 15–40% of patients respond
- Some develop autoimmune side effects: colitis, thyroiditis, pneumonitis
- Tumors can develop resistance mechanisms
- Access and cost remain issues worldwide

But the principle is proven: The immune system can be taught to see again.

And once it sees, the results can be breathtaking.

A New Philosophy of Healing

Checkpoint inhibitors didn't just give us new drugs.

They gave us a new faith in the body's ability to recognize, remember, and reclaim its power.

They taught us:

- That cancer is not invisible, just cleverly disguised
- That treatment can be enduring with immune memory lasting years
- That the best therapy may not be the strongest drug, but the smartest signal

"We didn't invent healing," said one immunologist.

"We removed what was blocking it."

Legacy and the Road Ahead

Checkpoint inhibitors now anchor standard-of-care regimens in oncology.

They've opened doors for:

- Neoantigen vaccines
- T-cell engagers
- Bi-specific antibodies
- Next-gen checkpoint targets beyond PD-1 and CTLA-4 (e.g., LAG-3, TIM-3, TIGIT)

The immune system, once a mystery, is now a medical partner.

Checkpoint inhibitors were the key.

The door they opened is still swinging wider every year.

"I Beat Cancer with my Own Immune System"

How the Body Remembered, Reawakened, and Reclaimed Life it was Never Meant to Lose

When Emma, age 38, heard the words "stage IV melanoma," she stopped breathing for a second.

She had two kids. A business. A life filled with to-do lists and normalcy.

She never expected a mole on her shoulder to become the thing that might take her away from all of it.

She asked the questions anyone would:

"How long do I have?"

"What are my options?"

"Is it too late?"

Her doctor surprised her with something new.

"Let's see what your immune system can do."

A New Kind of Treatment

Emma started pembrolizumab (Keytruda), an immunotherapy that unleashed her T-cells to recognize and destroy the cancer.

No hair loss.

No nausea.

Just a strange fatigue and the quiet wait for the first scan.

Weeks turned to months.

And then—news.

"The tumors are shrinking. Your body is fighting back."

Her immune system—previously blind to the cancer—had woken up.

What it Feels Like to Heal from the Inside Out

Unlike chemo, immunotherapy doesn't directly attack the tumor.

It reminds your immune cells what they forgot: This doesn't belong here. Destroy it.

Emma described the sensation not as a battle but as a reclamation.

"It felt like I was finally participating in my own healing. Like my body wasn't just being treated—it was rising."

Setbacks and Stillness

There were scary moments, too.

At month 4, she developed thyroid inflammation—a common immune-related side effect.

Her voice became hoarse. Her heart raced. She was pulled off treatment temporarily.

But even that was a sign:

Her immune system was awake, active, doing exactly what it was supposed to—with a bit too much enthusiasm.

Steroids helped. She recovered. Treatment resumed.

And the scans kept improving.

The Day She Heard the Words "No Evidence of Disease"

By month 12, the tumors were gone.

Not just smaller, undetectable.

She was still monitored, of course. Bloodwork. Imaging. The usual vigilance.

But the most important battle had already been won, not by a drug alone, but by the immune system that finally remembered her name.

What she says Now

Emma speaks at cancer survivorship events.

She mentors newly diagnosed patients.

She keeps the scan photos—the "before" and "after" of her internal revolution.

"My immune system didn't fail me. It just needed a key. Immunotherapy gave it that. And it did the rest."

She still lives with the shadow of cancer. But not with fear.

Because she knows her body is not defenseless.

It is vigilant.

One of Thousands and One of a Kind

Emma is not alone.

- Mark, with metastatic lung cancer, responded completely to nivolumab.
- Ravi, with advanced Hodgkin lymphoma, is now a medical student.
- Alina, once told she had 6 months, now runs marathons—and blood tests that show no trace of disease.

Each of them says a version of the same thing:

"I didn't just receive a drug. I got my immune system back. And it saved my life."

The Deeper Lesson

Immunotherapy isn't magic.

It's not always fast.

It doesn't work for everyone.

But when it does work, it taps into something ancient and profound:

- The body's own wisdom.
- Its buried strength.
- Its ability to distinguish between self and enemy.

"I didn't beat cancer with radiation. Or chemo.

I beat cancer with my own immune system.

And I will never stop being amazed by that."

Chapter 23

AI-Driven Drug Discovery

How Algorithms are Reshaping the Way we Invent Medicine—Faster, Smarter, and more Personal than Ever Before

For over a century, drug discovery relied on a mix of science, patience, and luck.

Penicillin was discovered by accident.

Statins were isolated from mold.

Cancer drugs were found through endless screening of chemical libraries.

The process was slow.

The odds of success? Dismal.

Out of 10,000 candidate molecules, maybe one becomes an approved drug.

And the journey takes 10–15 years and costs over $2 billion.

But now, a new kind of discovery engine has entered the lab.

Not a microscope.

Not a molecule.

But a machine learning model.

This is the era of AI-driven drug discovery—where computers help us predict, design, and optimize medicines in ways human minds alone never could.

The Problem AI Set Out to Solve

Drug development is a labyrinth of unknowns:

- Which targets matter?
- What molecular structures will bind them?
- Will it be safe in humans?

- Will it work across diverse populations?
- Can we find a molecule before the disease outpaces us?

Traditional methods are trial-and-error—slow, expensive, and often blind.

AI changes that by turning biology into data—and drug discovery into a problem solvable by code.

What AI does Differently

AI systems now assist at every stage of drug discovery:

- **Target Identification:** Machine learning analyzes genomic, proteomic, and disease pathway data to identify new biological targets—sometimes those human researchers overlooked.
- **Molecule Design:** AI models generate novel molecules from scratch—based on predicted binding to a target protein's 3D structure.
- **Prediction of Toxicity and Side Effects:** Before a molecule is ever synthesized, AI can simulate how it might affect the liver, heart, or brain—saving years of trial failures.
- **Personalization:** By combining genomic data with real-world patient outcomes, AI helps predict which drug will work best for which patient—a core tenet of precision medicine.

Real-World Milestones: The Future has begun

- **Insilico Medicine:** In 2021, Insilico used AI to design, validate, and begin testing a fibrosis drug in under 18 months—a process that normally takes up to 6 years.
- **Exscientia:** This UK-based company developed the first AI-designed molecule to enter human clinical trials—a candidate for the treatment of OCD.

They've also accelerated oncology, immunology, and inflammatory drug pipelines using reinforcement learning and deep generative models.

DeepMind's AlphaFold

Not a drug developer itself, but a revolution nonetheless.

AlphaFold predicted the 3D structures of nearly all known human proteins—a feat scientists expected to take decades. It has now become a foundational tool for researchers worldwide.

Speed, Cost, and Creativity

AI reduces:

- Time from target discovery to lead optimization
- Cost of failed compounds
- Waste in lab resources and animal models

But perhaps most exciting—it enables creativity without chemical bias.

AI models can explore entire regions of chemical space that human chemists never imagined.

As one researcher put it: “AI doesn’t just look where the light is. It looks everywhere.”

Human + Machine: A New Partnership

AI doesn’t replace scientists. It augments them.

Medicinal chemists, biologists, clinicians, and data scientists now work with:

- Neural networks
- Natural language processing models
- Predictive analytics
- Generative chemistry platforms

The result? A team that’s smarter, faster, and more precise than ever before.

Challenges and Cautions

This revolution isn’t without hurdles:

- Garbage in, garbage out: poor-quality data can mislead models
- Explainability: black-box predictions require human validation
- Clinical translation: not all promising predictions survive the complexity of real biology
- Ethics: privacy, bias, and AI ownership of intellectual property are unresolved questions

Still, the momentum is unmistakable.

What the Future Holds

In the coming decade, expect to see:

- AI-first pharma companies with end-to-end machine learning pipelines
- Predictive pharmacovigilance to catch adverse effects before rollout
- Adaptive clinical trials, guided by real-time AI learning
- And custom molecule generation tailored to an individual's genome, tumor profile, or microbiome

This isn't just a smarter way to make pills.

It's a smarter way to understand life.

"We used to wait for nature to reveal her secrets," one AI scientist said.

"Now, with AI, we're asking better questions. And getting answers faster than we ever dreamed."

From Molecules to Models to Miracles

AI isn't just speeding up discovery.

It's changing the very way we think about medicine—from reactive to predictive, from fixed to adaptive, from generic to exquisitely personal.

And as machines learn more about the patterns of disease, they may soon help us predict, prevent, and preempt illness itself.

The brain built silicon.

Now silicon is helping us heal the brain, the body, and the world.

Pharma Meets Algorithms: The Next Chapter

When Molecules Shake Hands with Machines, a New Kind of Healing Begins

For over a century, the pharmaceutical industry relied on chemists, biologists, and clinicians to find and refine the medicines that shaped human history.

Penicillin. Insulin. Chemotherapy. Statins.

These were discovered in petri dishes, flasks, and trial by fire.

But now, pharma is transforming again—not with a new class of drugs, but with a new kind of partner: Algorithms.

Not just as tools—but as co-discoverers, co-predictors, and co-pilots in the journey of healing.

We are entering the age of computational pharma—where biology meets big data, chemistry meets cloud, and every compound, trial, and decision is guided by intelligence that learns.

Data as a Drug Discovery Engine

Biology is no longer limited to what we can grow, test, or dissect.

Instead, it's becoming a field of high-dimensional data:

- Genomic sequences
- Proteomic landscapes
- Clinical trial outcomes
- Real-world patient data
- Imaging, sensor data, and voice biomarkers
- Social, behavioral, and environmental signals

Algorithms thrive in this space.

They find patterns humans miss.

They simulate entire biological systems.

They reduce 10 years of guesswork into 10 hours of computation.

And most crucially—they learn. The more data they get, the better they become.

The Next Chapter in Pharma: Smart, Self-Learning, Scalable

Here's how algorithms are reshaping every corner of pharmaceutical development:

- **Drug Discovery:** AI is designing molecules from scratch, predicting target fit, solubility, toxicity, and even synthetic feasibility—before they're ever made in a lab.
- **Personalized Therapies:** Machine learning matches the right drug to the right patient based on:
  - Genomic variations
  - Tumor profiles
  - Metabolic rates
  - Real-world adherence behavior
- **Clinical Trials**: Algorithms optimize trial design, predict dropout risks, identify responsive subgroups, and even simulate control arms using real-world data.
- **Global Health and Access:** AI helps map disease hotspots, optimize supply chains, and tailor public health interventions by analyzing population-level data at scale.

A New Culture of Collaboration

The pharma company of tomorrow will look more like:

- A biotech startup
- A data science lab
- A platform company
- A bioinformatics powerhouse

Biologists will work alongside AI engineers.

Regulatory teams will use dashboards powered by real-world evidence.

Medical affairs will use natural language models to parse feedback from thousands of HCPs and patients.

The lab bench now has a new neighbor: the neural network.

Pharma as Platform: From Products to Ecosystems

The industry is shifting from selling pills to delivering continuous, adaptive solutions:

- Drugs + digital therapeutics
- Devices that collect feedback and train the algorithm
- Companion apps that adjust dosage, flag side effects, and improve adherence
- Algorithms that learn from every patient and improve the next one's outcome

In this model, each patient helps the next.

Because every interaction trains the system.

The Ethical Frontier

When algorithms guide healthcare, we must ask:

- Who owns the data that trains the AI?
- How do we prevent bias from creeping in?
- Can we explain how decisions are made?
- How do we ensure access doesn't worsen inequality?

The answers will shape not just the future of pharma, but the trust it earns, or loses.

Responsible AI isn't just a goal. It's a prerequisite for progress.

From Reactive to Predictive, From Static to Living

The drug of the future won't be a static molecule.

It will be a living, learning system—capable of adjusting, evolving, and personalizing.

Imagine:

- A cancer therapy that learns your immune response and adapts in real time
- A mental health app that tunes digital therapy to your speech patterns and sleep

- A rare disease treatment that updates its dose based on live biomarker tracking

This isn't science fiction.

It's pharma's next chapter, and it's already being written in code.

A New Kind of Intelligence. A New Kind of Healing

The old pharma asked: What molecule treats this disease?

The new pharma asks:

What patterns explain this disease?

What combination of chemistry, code, and context will prevent it?

How can we design a health system that learns—patient by patient, year by year?

"In the past, we discovered drugs in the dark," says one AI–drug discovery pioneer.

"Now, the data lights the path. And the algorithm walks it with us."

Pharma Meets Algorithms and Medicine Awakens

We stand at the threshold of a future where medicine is:

- Faster to discover
- More precise to deliver
- Smarter as it learns
- And human-centered as never before

Because when pharma meets algorithms, it doesn't lose its soul.

It finds its future.

Early Breakthroughs and What's Coming

From First-Gen Algorithms to Next-Gen Medicines: The Evolution of a Revolution

The old model of drug discovery was measured in years, billions, and attrition rates.

Only 1 in 10,000 compounds reached the market.

Most failed in late-stage trials.

Costs were rising. Timelines were stalling. Hope, too often, came too late.

But then, something quietly disruptive emerged, not in a wet lab, but in the data center.

Artificial Intelligence didn't just promise to speed things up.

It offered a new language—one that could read biology, predict chemistry, and learn from every failed molecule ever tested.

This is the story of AI's first big wins—and the even bigger ones to come.

Early Breakthroughs: What AI has already Delivered

1. **AlphaFold: The Protein Puzzle Cracked**

   In 2021, DeepMind's AlphaFold predicted the 3D structures of nearly all known human proteins with stunning accuracy.

   What once took years in a lab could now be done in seconds by AI.

   This breakthrough has become a global open-access tool, unlocking insights in:

   - Structural biology
   - Enzyme engineering
   - Rare disease research
   - Antibody and vaccine development

   AlphaFold didn't just crack proteins. It cracked open the black box of molecular biology.

2. **Insilico Medicine: From Design to Trials in Record Time**

   Insilico used deep learning to:

   - Identify a novel target for pulmonary fibrosis
   - Generate lead compounds
   - Optimize them using reinforcement learning
   - And move a candidate drug into human clinical trials in 18 months

   This was one of the first full AI-to-clinic pipelines—a glimpse into a future where the computer doesn't just assist, but co-creates.

2. **Exscientia: The First AI-Designed Molecule in Humans**

   Exscientia, a UK-based firm, developed DSP-1181, a molecule targeting obsessive-compulsive disorder, in collaboration with Sumitomo Dainippon.

   From idea to preclinical candidate: less than 12 months.

   They've since moved multiple AI-designed oncology compounds into trials, including EXS21546, a selective A2A receptor antagonist.

4. **Atomwise: Predicting Drug-Target Interactions at Scale**

   Using convolutional neural networks, Atomwise's AI screens billions of molecules to find those most likely to bind to disease-relevant proteins.

   They've partnered with pharma giants and academia to develop:

   - Ebola antivirals
   - COVID-19 inhibitors
   - Targeted cancer agents

   AI is now scaling hit discovery, compressing timelines from years to days.

   The Impact so far

   - Design times slashed from 4–6 years to <1 year
   - Candidate libraries expanded from thousands to billions of virtual molecules
   - Failure prediction in toxicity and off-target effects improved before costly trials
   - Drug discovery democratized: biotech startups with AI platforms are now competing with Big Pharma

What's Coming: The Next Frontier in AI-Powered Pharma

1. **Generative Chemistry 2.0**

   Advanced models (like diffusion models and transformers) are now capable of:

   - Designing molecules with multiple constraints (efficacy + safety + synthetic feasibility)
   - Creating adaptive molecules that adjust based on live patient data
   - Exploring entire chemical spaces humans can't even imagine

   Soon, AI won't just suggest options. It will collaborate in real-time with chemists.

2. **Personalized Drug Discovery**

   With access to:

   - Genomic data
   - Electronic health records
   - Biomarker profiles

   ...AI will generate custom compounds based on an individual's biology.

   Imagine a cancer therapy designed not just for the tumor type, but for your tumor.

   This is precision pharma at its ultimate expression.

3. **AI-Powered Clinical Trials**

   AI will transform:

   - Trial design (adaptive arms, digital twins)
   - Patient recruitment (matching based on genetic and behavioral patterns)
   - Monitoring and prediction (detecting efficacy or toxicity earlier)

   This will make trials smaller, faster, cheaper, and more inclusive.

4. **AI + Synthetic Biology**

   The fusion of AI with DNA design and biofoundries will allow:

   - Automated creation of engineered cells and microbes
   - Rapid production of novel antibiotics, enzymes, and vaccines

- Smart delivery systems (e.g., AI-guided mRNA payloads or CRISPR delivery vectors)

5. **Closed-Loop Drug Discovery**

   Soon, AI models will be:

   - Continuously retrained on new trial data
   - Linked to real-world evidence from EHRs and wearables
   - Integrated into self-learning lab automation platforms

   This means: Every patient, every failure, every success makes the next drug smarter, faster, safer.

A New Paradigm: Pharma Becomes Predictive, Programmable, and Patient-Centric

In this next chapter, pharma isn't just reactive; it's anticipatory.

Instead of hoping a molecule works, AI predicts which one should work.

Instead of testing one-size-fits-all pills, AI helps create drugs that adapt to individual biologies.

Instead of 10 years to market, we may soon see candidate-to-clinic in 10 months.

A Word of Caution and Responsibility

This revolution must be guided by:

- Transparent AI models
- Ethical data sourcing
- Regulatory adaptability
- Global access, not just elite innovation

Because the algorithms are only as good as the values we build into them.

"The smartest lab in the world isn't a building. It's a network of models, minds, and molecules—learning in real time how to heal the human body."

Conclusion: From Algorithm to Apothecary

AI-driven drug discovery isn't just a tool.

It's a new lens through which we see biology—faster, deeper, and more creatively than ever.

The breakthroughs have already begun.

The next ones may redefine medicine entirely.

And the best part?

We've only just started training the models.

Chapter 24

Global Health Impact: Malaria, TB, and Neglected Diseases

Fighting the Unseen Killers with Science, Equity, and Relentless Determination

When we talk about the miracles of modern pharma, we often think of cancer cures or precision medicine.

But for billions around the world, the battle is more basic:

Surviving preventable, treatable, but historically neglected diseases.

Diseases that kill quietly.

That doesn't headline in rich countries.

That linger in the margins of poverty, displacement, and inequality.

Among them, none are more persistent—or more devastating—than malaria, tuberculosis, and the cluster of illnesses called neglected tropical diseases (NTDs).

This is the story of how pharma, global health organizations, and tireless frontline workers turned slow-burning tragedies into emerging triumphs.

Malaria: A Disease Older than Civilization

Malaria has stalked humanity for millennia.

- Caused by the Plasmodium parasite
- Transmitted by mosquitoes
- Killing children every two minutes, especially in sub-Saharan Africa

For decades, treatment relied on:

- Chloroquine (until resistance emerged)
- Then sulfadoxine-pyrimethamine
- And eventually, artemisinin-based combination therapies (ACTs)

The discovery of artemisinin (isolated from sweet wormwood by Chinese scientist Tu Youyou, who won the Nobel Prize) was a turning point.

ACTs became the gold standard supported by:

- Pharma partnerships (e.g., Novartis, Sanofi)
- Nonprofits like Medicines for Malaria Venture (MMV)
- Global procurement efforts like The Global Fund and Unitaid

Vaccines: The Next Weapon

In 2021, the WHO approved the world's first malaria vaccine—RTS,S (Mosquirix)—developed by GSK after 30+ years of research.

In 2023, a second, more scalable vaccine—R21/Matrix-M, developed by Oxford, was approved, promising wider access and greater affordability.

"For the first time, we can prevent infection before the mosquito ever bites."

The Results? Remarkable Progress

- Global malaria deaths fell by nearly 50% between 2000 and 2015
- Insecticide-treated nets, seasonal chemoprevention, and rapid diagnostics saved millions of lives
- Countries like Sri Lanka and China eliminated malaria entirely

But the fight isn't over:

- Climate change expands mosquito habitats
- Drug and insecticide resistance threaten gains
- Access remains fragile in conflict zones

Still, the direction is clear. We are winning a battle once thought unwinnable.

Tuberculosis: The World's Deadliest Curable Disease

TB kills over a million people each year, more than HIV or malaria.

It spreads silently, thrives in overcrowded settings, and disproportionately affects:

- The poor
- The immunocompromised
- Those with limited healthcare access

For decades, treatment was long, toxic, and incomplete.

But thanks to pharma-led R&D and global collaboration, breakthroughs have arrived.

Shorter, Safer Treatments

New drugs like:

- Bedaquiline (Janssen)
- Pretomanid (TB Alliance)
- Delamanid (Otsuka)

...have revolutionized treatment for drug-resistant TB, reducing therapy from 24 months to as little as 6–9 months, with fewer side effects.

These drugs are:

- Being produced at scale
- Distributed through public-private partnerships
- Saving lives in high-burden countries from India to South Africa

The TB Vaccine Pipeline

BCG—the current TB vaccine—is over 100 years old and only partially protective.

But a promising new candidate—M72/AS01E, developed by GSK and Aeras with Gates Foundation support—has shown nearly 50% efficacy in a major Phase II trial.

"A modern TB vaccine could change the trajectory of this disease forever."

Neglected Tropical Diseases (NTDs): A Quiet Revolution

NTDs include:

- Sleeping sickness
- Leishmaniasis
- Chagas disease
- Onchocerciasis (river blindness)
- Lymphatic filariasis
- Schistosomiasis
- And more

Together, they affect over 1.5 billion people, causing disability, stigma, and economic loss.

But here's what's remarkable:

Pharma didn't walk away.

They stepped up.

Pharma's Unprecedented Commitment

The London Declaration on NTDs (2012) brought together:

- WHO
- Governments
- Pharma companies (GSK, Merck, Novartis, Pfizer, J&J, Sanofi, Eisai)
- Nonprofits and donors

They pledged:

- Mass drug donations
- R&D for new treatments
- Logistics support for elimination programs

Impact at Scale

- Merck's ivermectin has helped reduce river blindness by over 90% in West Africa
- GSK has donated over 10 billion tablets of albendazole for lymphatic filariasis
- J&J, Pfizer, and Novartis have supported integrated control campaigns
- DNDi (Drugs for Neglected Diseases initiative) has developed new oral treatments for sleeping sickness and leishmaniasis

In 2020, 46 countries had eliminated at least one NTD.

This is global health at its best: Science + partnership + political will = measurable change.

A Future within Reach: What's Coming

- Next-gen vaccines for TB, malaria, and dengue
- Single-dose cures for NTDs
- AI-powered surveillance systems for outbreak prediction
- Long-acting injectables to ease delivery in remote areas
- mRNA platforms adapted for endemic diseases

And a growing commitment to health equity as not just a goal, but a moral obligation.

Final Reflection: Pharma's Greatest Victory?

It won't be the most profitable drug.

It may not win headlines.

It may never be used in the world's richest cities.

But eradicating malaria, curing TB, and eliminating NTDs would rank among pharma's greatest triumphs—because it would mean the industry didn't just heal the privileged. It stood for humanity itself.

"When we save a child from malaria, we don't just save a life—we save their future."

"When we cure TB, we don't just end an infection—we restore dignity and possibility."

"When we defeat neglected diseases, we prove that no disease—and no person—is too small to matter."

Novartis, GSK, and Public-Private Partnerships

When Global Pharma Collaborates for Global Good

There was a time when the pharmaceutical industry was seen as working alone—isolated in research labs, patent offices, and boardrooms.

But the world's greatest health challenges—malaria, tuberculosis, neglected diseases—don't respect boundaries, patents, or business models.

They demand something bigger.

And that's exactly what began to happen.

When Novartis, GSK, and other pharma leaders embraced public–private partnerships (PPPs), the world didn't just get better medicines.

It got better systems, better access, and better chances for millions who once had none.

This is the story of how these companies redefined what pharma can do—when it chooses to work with the world, not just sell to it.

GSK: A Legacy of Global Health Leadership

Few pharmaceutical companies have done more for neglected diseases and vaccine equity than GSK.

Tropical Disease Research and Product Development

- GSK's R&D hub in Tres Cantos, Spain, was established exclusively to develop medicines for neglected tropical diseases (NTDs) like malaria, leishmaniasis, and Chagas disease.
- The Tres Cantos Open Lab Foundation, launched in partnership with Wellcome and other global health groups, gives academic researchers access to GSK infrastructure—a unique collaborative model.

RTS,S (Mosquirix): The World's First Malaria Vaccine

- Developed by GSK in partnership with PATH and funded heavily by the Bill & Melinda Gates Foundation
- Took over 30 years and multiple generations of collaboration to develop

- Piloted in Ghana, Kenya, and Malawi with WHO support
- Now part of routine immunization programs, saving thousands of children's lives

Global Drug Donation Programs

- GSK has donated billions of doses of albendazole through WHO-led NTD programs for lymphatic filariasis and soil-transmitted helminths
- Collaborates with USAID, DFID, WHO, and numerous NGOs

"We know the value of our science grows when we share it," said one GSK executive. "Not just sell it."

Novartis: From Pharma Company to Global Health Architect

Novartis has redefined its role in low- and middle-income countries—not just as a supplier of medicines, but as a partner in building access and delivery systems.

The Novartis Malaria Initiative

- Over 1 billion treatments delivered since 1999—mostly at no profit
- Developed and scaled Coartem®, a WHO-recommended artemisinin-based combination therapy (ACT)
- Partnered with the Medicines for Malaria Venture (MMV) to develop Coartem Dispersible, the first ACT specifically formulated for children

NIBR (Novartis Institutes for BioMedical Research) in Global Health

- Novartis became one of the first Big Pharma companies to open a dedicated institute focused on non-commercial global health R&D
- Targets malaria, Chagas, cryptosporidiosis, leishmaniasis, and drug-resistant TB

Public-Private Collaboration in TB Treatment

- Developed bedaquiline-based regimens in partnership with Stop TB Partnership, the Global Fund, and UN agencies
- Supporting the 4TB initiative, aiming to eliminate TB in four high-burden countries through integrated care models

Novartis Access & Healthy Family Programs

- Novartis Access offers a portfolio of 15 essential medicines for NCDs in low-income countries at a cost of $1 per treatment per month

- Healthy Family Programs train local health educators and partner with ministries of health in countries like Kenya, Vietnam, and India

"Access isn't charity. It's a strategy," said a Novartis global health leader. "Because the world cannot be healthy unless everyone is."

Why Public–Private Partnerships Matter

These collaborations are more than feel-good philanthropy. They work because they combine:

- Pharma's science, scale, and supply chain
- Government and NGO reach
- Local trust and delivery
- Funding from global donors like the Gates Foundation, Unitaid, and The Global Fund
- Field data from patients and health systems

Together, they enable:

- Faster innovation
- Wider distribution
- Shared risk
- Sustainable impact

Real-World Impact: Shared Wins

1. **More Medicines to more People**

   PPPs helped deliver over 3 billion treatments for malaria and NTDs in the last decade.

2. **Better Vaccines for Vulnerable Populations**

   GSK, Pfizer, and others partner with Gavi, the Vaccine Alliance to deliver vaccines to over 800 million children.

3. **Accelerated R&D**

   GSK, Novartis, Sanofi, and Merck have shared compound libraries and research platforms with academic labs working on neglected diseases.

4. **Health System Strengthening**

   Novartis Access and Healthy Families train tens of thousands of frontline workers, creating lasting infrastructure.

What's Next: The Future of Pharma Collaboration

- AI-powered PPPs for faster vaccine and therapeutic development
- Shared manufacturing platforms for pandemic response
- Climate-resilient supply chains for medicines in fragile regions
- Global health equity funds jointly led by pharma, donors, and governments

These models will define how pharma works, not just for profit, but for purpose.

Final Thought: When Pharma Works With, Not Just for

Companies like GSK and Novartis are showing the way, not just in what they produce, but how they partner.

Because the future of pharma isn't about going it alone.

It's about standing shoulder to shoulder with governments, global organizations, and the communities most in need—to bring not just medicine, but justice.

"We believe in science," said one GSK researcher. "But more than that—we believe in sharing it."

"The real innovation," added a Novartis partner, "is what happens when we stop asking, 'Who owns this?' and start asking, 'Who needs this most?'"

Lives Saved in Millions across Africa and Asia

When Medicines Traveled Farther, Futures Grew Longer

Walk through a rural health post in Kenya, a neonatal ward in Bangladesh, or a mobile malaria clinic in Cambodia, and you'll see it: Survival.

Not just measured in data, but in cries of newborns, children playing with healed limbs, parents leaving with hope, not grief.

The quiet revolution in global health, led by access to life-saving medicines, has rewritten what it means to be born and grow up in the Global South.

Nowhere has this transformation been more visible, or more powerful than in Africa and Asia.

Malaria: Millions of Children who Lived

In the early 2000s, malaria was the leading killer of African children under five.

In response, the world mobilized:

- Artemisinin-based combination therapies (ACTs), developed and scaled by pharma partnerships
- Insecticide-treated bed nets
- Seasonal chemoprevention for at-risk regions
- Mass testing and community case management

Over 2 billion ACT treatments have been distributed—many through public–private efforts with Novartis, Sanofi, MMV, and global donors.

Lives Saved?

According to WHO and UNICEF, Over 11.7 million lives have been saved globally from malaria since 2000, most of them in Africa.

Vaccines: A Generation Protected

Through partnerships with Gavi, the Vaccine Alliance, pharma companies like GSK, Pfizer, Serum Institute of India, and others have delivered:

- Pentavalent vaccines (diphtheria, tetanus, pertussis, hepatitis B, Hib)
- Pneumococcal conjugate vaccines

- Rotavirus vaccines
- Meningitis A vaccines in the "meningitis belt" of Africa
- The world's first malaria vaccine, RTS,S

Impact?

- Over 1 billion children immunized since 2000
- More than 17 million deaths averted, per Gavi estimates

In Asia, vaccine rollout has transformed countries like India, Indonesia, Bangladesh, and Vietnam, driving down infant and under-five mortality at record speed.

HIV: From Death Sentence to Chronic Disease

In the early 2000s, HIV devastated sub-Saharan Africa.

Life expectancy plummeted. Mothers buried babies. Partners buried partners.

Then came a coordinated global effort:

- Generic ARTs, enabled by voluntary licensing and tiered pricing
- Pharma–NGO collaborations for pediatric formulations
- Community-based delivery models
- Prevention of mother-to-child transmission programs

Companies like GSK/ViiV, Cipla, and Mylan made life-saving antiretrovirals (ARVs) affordable and accessible.

Impact?

- UNAIDS estimates over 20 million lives saved globally due to ART rollout
- Over 28 million people currently on treatment, most of them in Africa and Asia

"I thought I would die in my twenties," says Nomsa from South Africa.

"Now I'm raising twins and running a sewing business."

Tuberculosis: A Slow but Steady Turnaround

- TB still kills more than 1.3 million people a year, but new tools are changing that.
- Short-course treatments with newer drugs (bedaquiline, delamanid)

- Digital adherence technologies in India
- GeneXpert rapid diagnostics across Africa
- Community DOTS programs

Pharma collaborations with the Stop TB Partnership, WHO, and national programs have scaled access.

Lives Saved?

- WHO estimates 74 million lives saved globally since 2000 due to TB care, most of them in Asia and Africa.

In India alone, where 25% of the world's TB burden resides, millions are now surviving each year who would once have been lost.

Maternal and Child Health: A Silent Surge in Survival

The expansion of access to:

- Oxytocin, magnesium sulfate (to prevent postpartum hemorrhage and eclampsia)
- Neonatal antibiotics and oral rehydration
- Zinc for diarrhea, iron, and folic acid for anemia
- Affordable contraceptives from public–private initiatives

...has transformed reproductive and child health.

Results?

- In sub-Saharan Africa, child mortality fell by over 60% between 1990 and 2020
- In South Asia, maternal mortality dropped by more than 65% in the same period

"My baby made it through pneumonia because the health worker had the right medicine," says Asha, a mother in rural Nepal.

"That was never guaranteed when I was a child."

Neglected Tropical Diseases: Quiet Killers, Now Defeated

Over 1.7 billion people are at risk of NTDs like lymphatic filariasis, river blindness, and soil-transmitted helminths.

But due to pharma-led donation programs, including:

- Merck's ivermectin

- GSK's albendazole
- J&J and Pfizer's azithromycin
- Novartis and Sanofi's antiparasitic R&D

...mass drug administration campaigns have reached hundreds of millions annually.

Impact?

- WHO estimates that over 40 countries have eliminated at least one NTD
- Millions spared disability, stigma, and blindness

A Legacy in Numbers and in Lives

Across Africa and Asia, the story is not just about drugs, but what those drugs made possible:

- Children reaching age five who wouldn't have
- Mothers surviving childbirth who once didn't
- Students staying in school without worm burden or malnutrition
- Communities free from fear of fevers, coughs, or skin ulcers

Each number in a health report is a name, a future, a chance reclaimed.

Final Reflection: What it Means to Save a Million Lives

When pharma works at scale, in collaboration with global health systems and communities, it does more than deliver molecules.

It shifts life expectancy curves.

It breaks generational cycles of disease and poverty.

It tells a different story about what's possible for the poorest on the planet.

"My child didn't die of malaria. My mother survived TB. I lived with HIV.

That's not just luck. That's the impact of science—shared equitably.

Epilogue

Why the World needs to Remember the Good Pharma

Because Memory Shapes Trust. And Trust Shapes the Future.

Pharma is often remembered for its missteps.

For the price tags, the scandals, the corporate battles.

And yes—those stories matter.

But there is another side to this story.

A side not told in headlines, but whispered in hospital wards, rural clinics, and family homes around the world.

A side shaped not by controversy, but by cure, comfort, and chance.

This book has been about that side.

The one where:

- A child with leukemia goes into remission
- A mother with HIV lives to raise her children
- A newborn is spared malaria
- A vaccine reaches the last mile
- A rare disease is no longer a life sentence

These are not miracles.

They are the result of science, risk, resolve—and yes, the pharmaceutical industry.

Why this Story must be Told

Because in the age of cynicism, we forget that:

- Pharma invented insulin, antibiotics, chemotherapy, and antiretrovirals

- Pharma helped eradicate smallpox, control polio, and contain Ebola
- Pharma turned HIV into a manageable condition
- Pharma created lifesaving drugs for children with cancer and babies with SMA
- Pharma supported global partnerships that delivered billions of treatments to the world's poorest

And while not perfect, pharma showed that profit and purpose can coexist—and sometimes, even align beautifully.

Because Gratitude should be Global

We remember to thank firefighters.

We remember to honor teachers.

We remember to revere frontline doctors.

We should also remember the unnamed scientists, chemists, data analysts, manufacturing teams, regulatory staff, and global logistics operators who—within pharma companies—worked quietly to save lives at scale.

Many of them gave their careers, their sleep, and in pandemic times, even their safety, to develop the tools that kept the world breathing.

Because the Future Depends on Balance

To hold the industry accountable is fair.

To demand equity is essential.

But to forget its triumphs is to distort history and cloud the future.

Because the next time we face a health crisis (and we will), the world will once again turn to:

- A researcher in a pharma lab
- A molecule waiting in development
- A factory preparing for scale
- A company ready to deliver at speed

If we want them to answer that call with urgency and trust, we must acknowledge that—when it mattered—they did.

Because This Isn't the End of the Story

The good pharma did is not confined to the past.

It continues:

- In AI-designed cancer drugs
- In one-time gene therapies
- In mRNA vaccines for HIV and RSV
- In long-acting malaria preventives
- In efforts to close the health equity gap for good

But the momentum of that good depends on belief.

And belief comes from truth.

Balanced truth. Whole truth. Remembered truth.

So let us remember pharma not as saint or sinner—but as a force that, when guided by science and service, has healed the world more than most will ever know.

Let us criticize when we must.

Push for fairness.

Demand accountability.

But let us also say clearly, and gratefully:

"You saved lives. You changed the story. You mattered."

The World needs to Remember—

Because the Good Pharma Did Is Still Saving Lives Today.

And Still Has So Much More to Give.

Post Script: Rebalancing the Narrative with Compassion and Facts

Not to Excuse. Not to Exalt. But to See Clearly—and Judge Fairly.

The pharmaceutical industry sits at a strange crossroads in the public imagination.

Revered in one breath. Reviled in the next.

Praised for saving lives. Criticized for pricing them.

Hailed as heroes in pandemics. Cast as villains in policy debates.

But beneath this tension is a deeper truth, a narrative out of balance.

And when narratives are unbalanced, so is public trust.

So is policy.

And so, ultimately, is progress.

This is not a call to silence criticism.

It is a call to temper outrage with insight, and memory with context.

The Danger of a One-Sided Story

We live in a time when:

- A single drug scandal defines a decade of research
- One pricing decision overshadows a thousand free drug donations
- The term "Big Pharma" conjures greed, not grit

And in that fog of cynicism, something precious is lost:

- The lives saved quietly in distant corners
- The breakthroughs born from patient, painful work
- The people—scientists, engineers, public health leaders—who give their lives to the pursuit of healing

We risk forgetting that pharma is not a monolith. It is a spectrum—flawed, yes, but also filled with brilliance and integrity.

Facts Matter

Here are some facts that rarely trend but always matter:

- Over 1 billion children have been vaccinated with pharma-developed products through Gavi
- More than 10 million people are alive today because of HIV medications
- Neglected Tropical Diseases are being eliminated in country after country, thanks to pharma donations
- 80% of essential medicines in low-income countries come through public–private collaboration
- And in pandemics, when every hour counts, pharma companies repurposed labs, accelerated trials, and built global supply chains in record time

These facts don't erase pharma's problems.

But ignoring them erases the full picture.

Compassion Matters Too

- Beyond the numbers are the stories:
- The immunologist who worked nights during COVID to optimize an mRNA formula
- The manufacturing line worker who ensured quality on a batch of lifesaving antibiotics
- The team that kept a cancer drug affordable in a low-income country even when the patent expired

Compassion means seeing that behind every brand name is a human being trying—often against odds—to make something that helps.

It also means recognizing the real-world barriers pharma faces:

- The risk of failed drugs
- The cost of trials
- The complexity of global access
- The burden of regulation, liability, and speed

To speak only in accusation is to flatten these efforts into caricature.

Why Rebalancing the Narrative Matters

Because how we see pharma shapes how we fund it, regulate it, partner with it, and rely on it.

If we only tell the worst of its story:

- We discourage innovation
- We erode the morale of the people trying to fix it
- We fail to hold constructive accountability

If we only tell the best:

- We risk complacency
- We ignore inequity
- We enable bad behavior

But when we balance both—fact with empathy, failure with progress, truth with proportion—we get the story we need:

A story that holds pharma to high standards and honors its high impact.

A story where we expect more because we've seen what's possible.

Let the narrative be honest.

Let it be complete.

Let it be driven not by outrage or obedience, but by a commitment to health, humanity, and truth.

Because Rebalancing the Narrative isn't just Fair..

It's the First Step to Building something Better.

The Human Faces behind Molecules

Because every Breakthrough Begins—and Ends—with someone's Story

When we talk about pharmaceutical breakthroughs, we speak in the language of science:

"First-in-class."

"Phase III efficacy."

"Multi-center, randomized trial."

"Regulatory approval."

"Molecular targets."

But every single one of those phrases hides a deeper truth:

Behind every molecule is a mosaic of people—often invisible, often unsung—whose choices, struggles, and dreams gave that medicine its chance to exist.

They are the human scaffolding behind our modern medicine cabinet.

And they deserve to be seen.

The Scientist in the Lab

She's not on the packaging.

You'll never know her name.

But she spent 15 years perfecting a delivery mechanism that allowed a fragile molecule to survive the stomach and reach the bloodstream intact.

She missed holidays.

She questioned herself.

She worked through failure after failure.

But she believed.

And because of her, a child with a rare genetic disorder now swallows a capsule every morning—and lives a full life.

The Trial Volunteer

He didn't sign up for glory.

He signed up because he lost a brother to a disease with no cure.

When the trial coordinator asked, "Are you sure?"

He said, "Yes. Someone has to try."

He endured side effects.

He traveled long distances to the clinic.

He answered every call, every questionnaire.

His blood, his response, his story helped validate the drug's safety.

You'll find his contribution in a footnote in the final paper—but without him, there's no therapy at all.

The Regulator and Reviewer

In a government office, a pharmacologist reviews the data.

Pages of adverse events, efficacy curves, confidence intervals.

This isn't red tape.

This is her moral responsibility.

She works late to ensure that what reaches the public deserves to.

She asks hard questions.

She looks for the blind spots.

She knows her approval means millions will trust the outcome.

And she carries that trust like a quiet burden.

The Factory Worker

He doesn't have a PhD.

But he calibrates machines with the precision of a surgeon.

He inspects vials by hand.

He counts by feel.

He knows when something is off—not from numbers, but from intuition honed by years on the floor.

He gets up before sunrise.

He doesn't get headlines.

But if he makes a mistake, a patient could suffer.

Because of him, quality lives not just in the formula, but in the finish.

The Patient Advocate

She's not from the pharma world.

She's from the world of loss.

Her child died waiting for a drug that didn't yet exist.

Now, she sits in boardrooms.

Pushes for trial accelerations.

Translates scientific jargon for frightened families.

Raises funds, awareness, and urgency.

Her voice isn't scientific. It's emotional. But that's what the system needed to remember.

Because she reminds everyone in the room:

"We're not talking about data. We're talking about someone's tomorrow."

The Doctor in the Field

He prescribes the drug in a refugee camp.

She administers it in a malaria-endemic village.

They explain it in dialects.

They improvise cold chains.

They carry hope in coolers.

They are the last mile in the journey of a molecule.

And without them, even the most brilliant drug goes nowhere.

The Parents who Waited. The Children who Lived

Behind every breakthrough are the ones who held their breath:

- The mother who carried a baby with a rare condition and waited for trial results
- The teenager with cystic fibrosis who dared to hope for something better

- The family in a rural village that lined up for a new vaccine
- The grandfather who said, "Let's try it" one last time—and lived to meet his great-grandchild

Their courage to believe in science gave pharma a chance to prove itself.

Why These Faces Matter

Because without them:

- Molecules stay on shelves
- Trials don't enroll
- Factories don't run
- Trust doesn't build
- Hope doesn't rise

Pharma doesn't happen in a vacuum.

It happens in humans.

And when we forget their stories, we reduce medicine to metrics.

The Truth?

There is no "miracle drug."

There is only miraculous effort—by hundreds, sometimes thousands—who bring that drug into being.

And if we truly want to honor medicine,

if we truly want to make it better,

we must start by remembering:

Every life-saving dose began with someone who dared to try, dared to wait, dared to believe.

Behind every molecule is a human face.

Look closer. The future of medicine is made there.

The Need for a Constructive, Responsible, and Grateful Engagement with Pharma

Because Change Comes not from Conflict or Compliance—But from Collaboration and Clarity

The pharmaceutical industry touches nearly every life on earth.

It develops the medicines we take.

It drives the vaccines we receive.

It funds much of the biomedical research that shapes global health.

And yet, the relationship between society and pharma is often fraught.

Public discourse swings between distrust and dependence, anger and admiration.

But there is a better way.

A way rooted not in passive acceptance or aggressive opposition, but in thoughtful, responsible, and ultimately grateful engagement.

Because the future of medicine depends not just on what pharma does—

But on how we, as a global society, choose to engage with it.

1. **Why Constructive Engagement Matters**

   Constructive engagement is about asking:

   How can we help pharma do better, while recognizing what it already does well?

   It means:

   - Encouraging innovation without tolerating abuse
   - Supporting access without undermining sustainability
   - Partnering on public health goals without silencing legitimate critique

   It's not about being naïve.

   It's about being informed and intentional—because yelling from the outside doesn't build better systems. Working together does.

2. **Why Responsibility Matters**

   We, as citizens, policymakers, academics, and advocates, have a responsibility too:

   - To understand the complexity of drug development and regulation
   - To demand transparency in pricing and decision-making
   - To hold pharma accountable for ethical lapses
   - To champion equity and fairness in access and research priorities

   But responsibility also means:

   - Recognizing that zero-risk, zero-cost, zero-profit expectations are unrealistic
   - Accepting that scientific discovery is uncertain, messy, and expensive
   - Supporting the systems—like IP protections and global funding coalitions—that make innovation and access possible

   In other words, we must expect more and know more.

3. **Why Gratitude still has a Place**

   Gratitude is not complicity.

   It's not a surrender of scrutiny.

   Gratitude is an acknowledgment:

   - That antibiotics saved millions before we understood their mechanism
   - That ARTs gave back decades of life to those with HIV
   - That vaccines ended smallpox and turned polio into a rarity
   - That even during a pandemic, pharma moved at warp speed—not perfectly, but purposefully

   We can critique the system while appreciating its successes.

   We can push for better while saying, sincerely: Thank you for what you've done right.

   Because that gratitude fosters something more powerful than outrage:

   A will to improve together.

4. **What this Relationship can become**

   If we recalibrate this relationship, we can build:

   - Faster response to pandemics
   - Smarter regulation and R&D funding

- More ethical innovation
- Broader access to lifesaving therapies
- Health systems that learn, adapt, and serve—globally

But that takes collaborative effort, not conflict.

Shared responsibility, not blame-shifting.

Hope anchored in realism, not ideology.

Final Reflection: A Call to All Stakeholders

To journalists: Cover pharma's failures, but also its milestones.

To activists: Demand access—but also recognize what it takes to make a drug exist.

To regulators: Be firm, but not adversarial.

To pharma leaders: Be bold, but transparent.

To citizens: Ask questions. Stay curious. Stay engaged.

Because this is not an industry to dismantle or deify.

It's one to challenge, improve, and co-create with—because it's here to stay.

And because, ultimately, when we collaborate constructively,

Patients win. Science thrives. And society moves forward—together.

Let our engagement be informed.

Let our expectations be high.

Let our gratitude be real.

Let our partnership with pharma be not perfect—but purposeful.

APPENDICES

Appendix A: A Timeline of Major Breakthrough Drugs

Appendix B: Index of Diseases and Breakthrough Drugs

Appendix C: Pharma's Global Footprint: Key Milestones by Region

Appendix D: Honoring the Scientists behind the Molecules

Appendix E: Key Public-Private Collaborations

Appendix A

A Timeline of Major Breakthrough Drugs

Part I: The Foundations of Healing (Pre-World War Era) Aspirin (1899)

- **Why it mattered**: The first true mass-market drug, providing safe, reliable pain relief and anti-inflammatory benefits.
- **Breakthrough element**: Transition from herbal remedies (willow bark) to chemically synthesized, standardized medicine.
- **Impact**: Millions gained relief from pain and fevers; it became the world's most widely used drug and a foundation for modern pharmacology.

Insulin (1921)

- **Why it mattered**: Transformed diabetes from a fatal disease into a manageable chronic condition.
- **Breakthrough element**: First hormone therapy successfully used in humans.
- **Impact**: Saved countless lives, with early patients dramatically rescued from near-death conditions. A cornerstone of chronic disease management.

Sulfa Drugs (1930s)

- **Why it mattered**: The first effective antibacterial agents before penicillin.
- **Breakthrough element**: Derived from dyes, showing that chemicals could kill bacteria selectively.
- **Impact**: Reduced deaths from pneumonia, sepsis, and childhood infections—paving the way for the antibiotic revolution.

Part II: Post-War Pharmaceutical Revolution (1940s–1960s)

Penicillin (1940s)

- **Why it mattered:** The first modern antibiotic saved soldiers and civilians alike.
- **Breakthrough element:** Mass production during WWII—ushering in the "antibiotic era."
- **Impact:** Mortality from infections plummeted; surgical and medical practices transformed.

Streptomycin (1943)

- **Why it mattered**: The first effective treatment for tuberculosis (then a leading cause of death).
- **Breakthrough element**: Soil-derived antibiotic.
- **Impact**: Helped empty TB sanatoriums and gave patients back their lives.

Cortisone (late 1940s)

- **Why it mattered**: Revolutionized treatment of inflammatory and autoimmune diseases.
- **Breakthrough element**: First synthetic corticosteroid.
- **Impact**: Allowed arthritis patients to regain mobility; opened the door for managing asthma, allergies, and lupus.

Polio Vaccine (1950s)

- **Why it mattered**: Ended widespread fear of paralysis and epidemics.
- **Breakthrough element**: First mass vaccination campaigns.
- **Impact**: Millions of children saved; polio nearly eradicated worldwide.

Part III: The Golden Age of Drug Discovery (1970s–1990s)

Beta-Blockers

- **Why it mattered**: Reduced risk of sudden cardiac death and heart attacks.
- **Breakthrough element**: First rationally designed cardiovascular drugs (James Black's work).
- **Impact**: Millions of lives prolonged, establishing "preventive cardiology."

SSRIs (e.g., Prozac)

- **Why it mattered**: Transformed psychiatry by providing effective, safer antidepressants.
- **Breakthrough element**: Shift from stigma and institutionalization to outpatient, patient-centered care.
- **Impact**: Millions relieved from depression; reshaped mental health treatment.

Statins (e.g., Lipitor)

- **Why it mattered:** Revolutionized cholesterol management, reducing heart attacks and strokes.
- **Breakthrough element:** Discovery that molds could lower cholesterol synthesis.
- **Impact:** Extended millions of lives, redefining cardiovascular risk management.

ACE Inhibitors

- **Why it mattered**: Tackled hypertension and heart failure effectively.
- **Breakthrough element**: Derived from snake venom peptides.
- **Impact**: Transformed outcomes for generations with high blood pressure.

IVF and Hormonal Therapies

- **Why it mattered**: Opened doors to parenthood for infertile couples.
- **Breakthrough element**: Clomiphene, gonadotropins, and IVF protocols.
- **Impact**: Millions of births; emotional and social transformations worldwide.

Part IV: Pharma in the Face of Crisis (1980s–2000s)

AZT (1987)

- **Why it mattered**: First antiretroviral drug for HIV/AIDS.
- **Breakthrough element**: Repurposed anticancer compound.
- **Impact**: Gave hope in the darkest days of the epidemic.

HAART (Triple Therapy, 1996)

- **Why it mattered:** Turned HIV from a death sentence into a chronic condition.
- **Breakthrough element**: Combination therapy halted viral replication.
- **Impact**: Extended life expectancy for millions; changed global health landscapes.

Gleevec (2001)

- **Why it mattered**: The first targeted cancer therapy.
- **Breakthrough element**: Specifically blocked the abnormal protein driving CML.
- **Impact**: Transformed certain cancers from lethal to manageable.

Expanded Vaccines

- **Why it mattered**: Hepatitis B, HPV, and rotavirus vaccines prevented millions of deaths quietly.
- **Impact**: Protected future generations from cancer and infections.

Part V: Biologics and Personalized Medicine (2000s–2020s)

Monoclonal Antibodies (Rituximab, Herceptin, Keytruda)

- **Why it mattered**: A new era in cancer and autoimmune therapy.
- **Breakthrough element**: Harnessed immune system precision.
- **Impact**: Patients with lymphoma, breast cancer, and melanoma gained hope and cures.

Hepatitis C Cures (Sofosbuvir, 2013)

- **Why it mattered**: First complete cure for a chronic viral infection.
- **Impact**. Millions spared cirrhosis and liver cancer; a true redemption story.

Orphan/Rare Disease Drugs (Spinraza)

- **Why it mattered**: Addressed conditions once ignored.
- **Impact**: Children with SMA could walk, live, and thrive.

COVID-19 Vaccines (mRNA, 2020)

- **Why it mattered**: Humanity's fastest large-scale medical triumph.
- **Breakthrough element**: mRNA technology.
- **Impact**: Saved millions, proving the power of science under crisis.

Part VI: The Future

Gene Therapy (Luxturna, Zolgensma)

- **Why it matters**: Restores sight, cures genetic paralysis in infants.
- **Impact**: Turning inherited tragedy into triumph.

CRISPR and Immunotherapies

- **Why it matters**: Redefining cures for genetic and cancer diseases.
- **Impact**: Moves medicine toward true personalization and self-healing.

In essence: Each of these drugs marks not just a scientific advance, but a profound reshaping of human destiny—transforming how long and how well we live.

Appendix B

Index of Diseases and Breakthrough Drugs

Disease/Condition	Breakthrough Drug(s)	Year	Impact
Chronic Pain	Aspirin	1899	First modern analgesic
Epilepsy	Pehnobarbital, Phenytoin	1912 1938	First effective anti-seizure drugs
Diabetes	Insulin	1921	Life-saving for Type 1 diabetes
Pneumonia / Sepsis	Sulfa Drugs	1930s	First antibacterial agents
Infections	Penicillin	1940s	Revolutionized infection care
Tuberculosis	Streptomycin Multi-drug Regimens	1943 1950s onward	First effective anti-TB drug Standardized TB cure globally
Malaria	Chloroquine Artemisinin Mosquirix Vaccine	1940s 1970s 2019-21	Standard malaria therapy Modern gold-standard treatment First malaria vaccine deployed
Arthritis / Inflammation	Cortisone	1949	First synthetic steriod; restored mobility
Asthma / Allergies	Corticosteroids	1950s	Revolutionized chronic asthma management
Diabetes	Metformin	1950s	Foundation for Type 2 diabetes
Mental Health	Chlorpromazine	1950s	First antipsychotic, deinstitutionalization
Polio	Polio Salk Vaccine Sabin Oral Vaccine	1955 1960s	First inactivated polio vaccine Simplified global immunization
Infertility	Clomiphene	1960s 1978	Enabled ovulation induction

Contd..

Disease/Condition	Breakthrough Drug(s)	Year	Impact
	IVF Protocol Drugs		Supported assisted reproduction
NSAIDs	Ibuprofen, Naproxen	1960s	Safer, accessible pain control
Biploar Disorder	Lithium	1970s	Stabilized biploar disorder
Cardiovascular Disease	Beta-Blockers	1970s	Reduced sudden cardiac deaths
Hypertension	ACE Inhibitors	1980s	Global standard in BP control
Depression	SSRI (Prozac)	1980s	Transformed psychiatry, broke stigma
Hepatitis B	Hepatitis B Vaccine	1981	First anti-cancer preventive vaccine
HIV / AIDS	AZT	1987	First anti-retroviral
Hypercholestremia	Statins	1990s	Revolutionized cholesterol management
HAART	Triple therapy	1996	Turned HIV into chronic disease
Diabetes / Obesity	GLP-1 Agonists	2000s	Dual glucose & weight control
Vaccine	Rotavirus Vaccine	2006	Prevented childhood diarrheal deaths
Cervical Cancer	HPV Vaccine	2006	Prevents cervical cancer
Hepatitis C	Sofosbuvir	2013	Curative therapy for chronic viral disease
Rare Diseases (SMA)	Spinraza	2016	First SMA therapy
Rare Diseases (Blindness)	Luxturna	2017	First approved gene therapy for sight restoration
SMA in Infants	Zolgensma	2019	Gene therapy cure in infants
Viral Epidemics	COVID-19 Vaccines (mRNA)	2020	First mRNA vaccine in crisis response

Appendix C

Pharma's Global Footprint: Key Milestones by Region

Pharmaceutical innovation has never been the achievement of a single nation. It is a global tapestry woven by scientists, doctors, entrepreneurs, policymakers, and patients across continents. Each region has played a vital role, contributing not only breakthrough molecules but also new models of access, affordability, and care delivery.

1. **Europe: The Cradle of Modern Pharma**

   Germany (late 1800s – early 1900s)

   The birthplace of modern pharmaceutical chemistry. Bayer's discovery of Aspirin (1899) became the world's first blockbuster drug. German firms like Hoechst and Merck pioneered industrial-scale drug manufacturing, turning chemistry into healing.

   Impact: Europe defined the model of industrial pharmaceutical R&D.

   United Kingdom (1920s–1950s)

   - Insulin mass production (1923, in collaboration with Canada).
   - Penicillin scaling during WWII (Florey & Chain, Oxford).
   - **Impact**: Life-saving drugs became available globally because British scientists solved the problem of scale.

   France & Switzerland (20th Century)

   Roche led in vitamins and later psychopharmaceuticals.

   Sanofi emerged as a vaccine powerhouse.

   - **Impact**: European pharma diversified into psychiatry, nutrition, and vaccines—fields that remain critical today.

2. **North America: The Era of Scale and Innovation**

   United States (1940s–present)

   - World War II catalyzed the mass production of penicillin.
   - NIH, FDA, and clinical trial infrastructure made the U.S. a global R&D hub.
   - Blockbusters like statins, SSRIs, ACE inhibitors, and cancer therapies were commercialized here.
   - The biotech revolution (Genentech, Amgen, Biogen) in the 1970s–80s created a new era of biologics.
   - COVID-19 vaccines (Pfizer/BioNTech, Moderna) validated decades of investment in molecular biology and mRNA science.
   - Impact: The U.S. became the engine of pharmaceutical innovation, often setting global regulatory and clinical standards.

3. **Canada (1921)**

   - Discovery of insulin by Banting and Best in Toronto.
   - **Impact**: A miracle that transformed diabetes into a manageable condition.

4. **Asia: From Traditional Remedies to Global Powerhouses**

   Japan (1950s–1980s)

   - Pioneered antibiotics and cardiovascular drugs post-WWII.
   - Companies like Takeda, Daiichi Sankyo, and Astellas became global innovators.
   - **Impact**: Japan blended traditional medicine with advanced R&D, leading in cardiovascular and oncology research.

   India (1970s–present)

   - Post-1970 patent law reforms enabled the rise of the generic drug industry.
   - Became the "Pharmacy of the World," supplying affordable ARVs for HIV/AIDS and essential generics worldwide.
   - Major vaccine producer (Serum Institute of India), including during the COVID-19 crisis.
   - **Impact**: Democratized access to life-saving medicines for millions in low- and middle-income countries.

China (2000s–present)

- Major investments in biotech, vaccines, and the modernization of traditional medicine.
- Now a leader in biosimilars, gene therapies, and AI-driven drug discovery.
- **Impact**: China is positioning itself as a future giant of innovation and manufacturing scale.

4. **Latin America: Regional Pioneers and Public Health Models**

Cuba (1980s–2000s)

- Known for biotech innovation despite economic constraints, including vaccines and monoclonal antibodies.
- **Impact**: Demonstrated how small nations can lead in scientific niches.

Brazil & Mexico

- Expanded vaccine production and distribution systems, partnering with the WHO and Gavi.
- Brazil's public health system leveraged generics for broad HIV treatment programs.
- **Impact**: Regional leadership in integrating pharma with public health strategies.

5. **Africa: From Dependency to Emerging Capacity**

South Africa (2000s–present)

- Central in the fight for affordable HIV/AIDS drugs, challenging global pharma on pricing.
- Recently invested in mRNA vaccine manufacturing capacity.
- **Impact**: A catalyst for discussions on global equity in pharma.

Continental Milestones:

- Expanded partnerships with Novartis, GSK, and WHO for malaria and TB drug access.
- **Impact:** Millions of lives saved through coordinated public–private initiatives.

6. **Middle East: Building R&D Bridges**

   Israel

   - Strong biotech and generics industry, with Teva Pharmaceuticals as a global leader in generic drugs.
   - **Impact**: Critical supplier of affordable medicines and a hub for biotech startups.

   Gulf States

   - Heavy investments in pharma manufacturing and clinical research in recent decades.
   - **Impact**: Emerging as regional healthcare innovation hubs.

Conclusion: A Shared Global Legacy

The pharmaceutical story is not just Western, Eastern, Northern, or Southern—it is universal. Each region contributed:

- Europe gave chemistry and structure.
- North America scaled innovation and clinical rigor.
- Asia democratized access and affordability.
- Latin America and Africa pioneered public health models and affordability battles.
- The Middle East bridged biotech and generics.

Together, these milestones remind us: pharma's footprint is global, its destiny shared, and its story still unfolding.

Appendix D

Honoring Scientists behind the Molecules

Celebrating the Pioneers who Changed the Course of Medicine

Behind every breakthrough drug, there is a scientist—or a team of them—who asked, "What if?"

They worked in obscurity, failed often, and sometimes waited decades for recognition.

But their ideas, their experiments, and their perseverance helped transform the future of health.

This appendix honors just a few of the many whose work built the foundation for the modern pharmacopeia.

Emil von Behring

- **Breakthrough**: Diphtheria antitoxin (1890s)
- **Contribution**: Developed the first effective therapy based on immune serum—laying the foundation for immunology and biologics.
- **Legacy**: First Nobel Prize in Medicine (1901)

Tu Youyou

- **Breakthrough**: Artemisinin (1972)
- **Contribution**: Extracted artemisinin from sweet wormwood using ancient Chinese texts, leading to the most important modern malaria treatment.
- **Legacy**: Nobel Prize (2015); saved millions from malaria.

Alexander Fleming

- **Breakthrough**: Penicillin (1928)
- **Contribution**: Discovered mold with bacteria-killing properties; inspired the antibiotic revolution.
- **Legacy**: Shared Nobel Prize in 1945 with Florey and Chain.

Frederick Banting & Charles Best

- **Breakthrough**: Insulin (1921)
- **Contribution**: Isolated insulin from pancreatic extracts, turning diabetes from fatal to treatable.
- **Legacy**: Banting received the Nobel Prize in 1923 at age 32.

Gertrude Elion & George H. Hitchings

- **Breakthroughs**: Mercaptopurine, acyclovir, azathioprine, and early antivirals (1940s–1980s)
- **Contribution**: Rational drug design without animal testing; developed the precursors to modern chemotherapy and antiviral therapies.
- **Legacy**: Nobel Prize in Medicine (1988); paved the way for precision medicine.

Jonas Salk

- **Breakthrough**: Inactivated polio vaccine (1955)
- **Contribution**: Developed the first safe and effective vaccine against paralytic polio.
- **Legacy**: Refused to patent the vaccine—"Could you patent the sun?"

James Black

- **Breakthroughs**: Propranolol (beta-blockers), Cimetidine (H2 blocker)
- **Contribution**: Revolutionized the treatment of hypertension and peptic ulcers through receptor-targeted drugs.
- **Legacy**: Nobel Prize (1988); father of modern pharmacology.

Paul Janssen

- **Breakthroughs:** Over 80 drugs, including haloperidol, fentanyl, and loperamide
- **Contribution:** Founded Janssen Pharmaceutica; prolific drug developer in psychiatry, pain, and infectious diseases.
- **Legacy:** His legacy continues through Johnson & Johnson's pharmaceutical arm.

Jennifer Doudna & Emmanuelle Charpentier

- **Breakthrough**: CRISPR-Cas9 gene editing

- **Contribution**: Developed a precise method to edit human DNA, opening the door to curative gene therapies.
- **Legacy**: Nobel Prize (2020); foundational to next-gen drug discovery.

Katalin Karikó & Drew Weissman

- **Breakthrough**: mRNA vaccine platform (used in COVID-19 vaccines)
- **Contribution**: Their decades of research enabled the rapid development of COVID-19 vaccines using synthetic mRNA.
- **Legacy**: Nobel Prize (2023); sparked a new era in vaccinology.

Brian Druker

- **Breakthrough**: Imatinib (Gleevec)
- **Contribution**: Transformed chronic myeloid leukemia into a manageable condition; pioneer of targeted therapy.
- **Legacy**: A model for precision oncology.

Frank Duffin and Stephen O'Brien

- **Breakthrough**: Sofosbuvir (Sovaldi)
- **Contribution**: Developed the molecule that led to the first near-certain cure for Hepatitis C.
- **Legacy**: Revolutionized antiviral therapy with short, safe oral treatment.

Why it Matters

These individuals and teams did more than discover molecules.

They:

- Transformed disease outcomes
- Inspired generations of researchers
- Proved that vision and persistence can save lives

Some were honored with prizes. Many were not.

All contributed to a better, longer, healthier life for humanity.

Appendix E

Key Public-Private Collaborations

Early 20th Century Foundations

- 1921–1923 | University of Toronto + Eli Lilly (Insulin Commercialization)
- **Collaboration**: Banting & Best's discovery of insulin was scaled globally through Eli Lilly's production.
- **Impact**: Diabetes transformed from a death sentence into a manageable condition.

1940s | U.S. & U.K. Governments + Pharma (Penicillin WWII Project)

- **Collaboration**: Government labs, Pfizer, Merck, and others mass-produced penicillin for soldiers.
- **Impact**: Infection mortality plummeted during WWII, sparking the modern antibiotic era.

Post-War Expansion (1950s–1970s)

- 1950s | Salk Institute + U.S. Public Health + Foundations (Polio Vaccine)
- **Collaboration**: Jonas Salk's work was supported by the March of Dimes, and mass immunization campaigns were publicly funded.
- **Impact**: Polio incidence fell dramatically, leading toward global eradication.

1960s–1970s | WHO + Pharma (Smallpox Eradication Campaign)

- **Collaboration:** WHO coordinated vaccine donations and distribution with producers.
- **Impact:** Smallpox eradicated in 1980—the single greatest victory of vaccination.

Golden Age of Global Health Partnerships (1980s–2000s)

1987 | Burroughs Wellcome + FDA + Activist Groups (AZT for HIV/AIDS)

- **Collaboration**: First accelerated drug approval process under activist pressure and pharma–regulator cooperation.
- **Impact**: Gave HIV patients a lifeline; transformed drug approval culture.

1990s | WHO + Pharma + Global Health NGOs (Expanded Program on Immunization)

- **Collaboration**: UN agencies, Gavi's precursor alliances, and companies like Merck and GSK expanded the routine childhood vaccination program.
- **Impact**: Massive declines in measles, diphtheria, and tetanus.

1996 | Pharma + Activists + Governments (HAART Rollout)

- **Collaboration**: Triple therapy development + global HIV access campaigns.
- **Impact**: Turned HIV into a chronic condition in high-income countries, setting stage for global rollout.

2000 | Gavi, the Vaccine Alliance (Bill & Melinda Gates Foundation + WHO + UNICEF + Pharma)

- **Collaboration**: Created to improve vaccine access in low-income nations.
- **Impact**: Vaccines for Hepatitis B, Hib, rotavirus, and HPV reached millions in Africa and Asia.

21st Century Breakthrough Collaborations

2001 | Novartis + Academic Hospitals (Gleevec Trials)

- **Collaboration**: Pharma + academia partnered to pioneer targeted therapy.
- **Impact**: First precision oncology breakthrough, redefining cancer treatment.

2006 | PATH + GSK + Bill & Melinda Gates Foundation (Malaria Vaccine Development)

- **Collaboration**: Decades-long effort that produced RTS,S (Mosquirix), the first malaria vaccine.
- **Impact**: Introduced into African immunization programs in 2019–2021.

2010 | DNDi (Drugs for Neglected Diseases initiative)

- **Collaboration:** A public–private nonprofit with the WHO, pharma companies, and NGOs.
- **Impact:** Delivered affordable treatments for sleeping sickness, leishmaniasis, and Chagas disease.

2013 | Gilead + Global Health Orgs (Hepatitis C Licensing Agreements)

- **Collaboration**: Voluntary licenses with generic manufacturers in India enabled low-cost cures.
- **Impact**: Millions in LMICs gained access to hepatitis C eradication.

COVID-19 Era and Beyond (2020s)

2020 | BioNTech + Pfizer + Global Regulators (mRNA Vaccines)

- **Collaboration**: First mRNA vaccine developed at record speed through multinational partnerships.
- **Impact**: Saved millions of lives; established mRNA as a new medical platform.

2020 | AstraZeneca + Oxford University + Serum Institute of India + Gavi (COVAX Initiative)

- **Collaboration**: Ensured billions of doses for low- and middle-income countries.
- **Impact**: Enabled broad global vaccine access during the pandemic.

2021 | Moderna + NIH + U.S. Operation Warp Speed

- **Collaboration**: The U.S. government provided funding and infrastructure for the mRNA rollout.
- **Impact**: Validated public–private emergency innovation at scale.

2022 | Africa CDC + BioNTech + WHO (African mRNA Hubs)

- **Collaboration**: Technology transfer partnerships to establish vaccine manufacturing in Africa.
- **Impact**: First step toward continent-wide self-reliance in vaccines.

Conclusion

From insulin in Toronto to mRNA hubs in Africa, pharma's greatest triumphs have often come not through solo effort but through public–private partnerships: governments providing vision and funding, academia offering discovery, NGOs pushing equity, and pharma scaling solutions. Together, these collaborations not only created breakthrough drugs but also changed the destiny of global health.

Glossary of Key Terms

A–C

ACT (Artemisinin-based Combination Therapy)

A gold-standard treatment for malaria combining artemisinin with another antimalarial to reduce resistance.

Adjuvant

A substance added to vaccines to enhance the body's immune response to an antigen.

Antibody (Monoclonal)

Lab-engineered proteins that bind to specific targets (like cancer cells) to treat disease with high precision.

Antiretroviral Therapy (ART)

Treatment regimen for HIV that suppresses viral replication and prevents disease progression.

Biosimilar

A biologic product highly similar to an already approved biologic, with no clinically meaningful differences in safety or effectiveness.

Breakthrough Therapy Designation

A U.S. FDA program that expedites development of drugs for serious conditions showing substantial improvement over existing therapies.

CRISPR (Clustered Regularly Interspaced Short Palindromic Repeats)

A revolutionary gene-editing technology that allows precise modification of DNA in living organisms.

D–G

Drug Repurposing

The strategy of using existing drugs to treat new or different diseases (e.g., sildenafil was originally studied for angina, repurposed for ED).

EMA (European Medicines Agency)

The regulatory authority that evaluates and approves medicines for use across the European Union.

First-in-Class

A drug that uses a new and unique mechanism of action for treating a medical condition.

Gene Therapy

Treatment that involves modifying or replacing defective genes to treat or cure a disease at the genetic level.

Generics

Drugs that are chemically identical to branded versions but sold under different names, often after patent expiry.

GMP (Good Manufacturing Practice)

A system ensuring that drugs are consistently produced and controlled according to quality standards.

H–M

Humanized Monoclonal Antibody

An antibody engineered to closely resemble human immune proteins to reduce immune reactions when used as a therapy.

HTA (Health Technology Assessment)

A multidisciplinary evaluation of the medical, economic, social, and ethical implications of health technologies.

Immunotherapy

Treatments that stimulate or harness the immune system to fight diseases like cancer and autoimmune conditions.

In Vivo / In Vitro

In vivo = within a living organism; in vitro = in a lab environment (e.g., test tube or petri dish).

mRNA Vaccine

A type of vaccine that uses messenger RNA to instruct cells to produce a protein that triggers an immune response.

Mechanism of Action (MoA)

The specific biochemical interaction through which a drug produces its therapeutic effect.

N–R

Neglected Tropical Diseases (NTDs)

A group of infectious diseases that primarily affect low-income populations in tropical and subtropical regions.

Novel Chemical Entity (NCE)

A compound that has not been previously approved or marketed in any form for medical use.

Orphan Drug

A medication developed to treat rare diseases affecting fewer than 200,000 people in the U.S. (or equivalent thresholds elsewhere).

Pathogen

Any organism—such as bacteria, viruses, or parasites—that causes disease.

Personalized Medicine

Medical care tailored to the individual characteristics, genetics, and biology of each patient.

Pharmacovigilance

The science and activities related to detecting, assessing, and preventing adverse drug effects.

S–Z

Small Molecule Drug

Traditional drugs composed of low molecular weight compounds that enter cells easily and interact with intracellular targets.

Sustained-Release / Long-Acting Formulation

Drug delivery systems designed to release medication slowly over time to maintain therapeutic levels.

Therapeutic Index

A measure of a drug's safety—comparing the amount that causes a therapeutic effect to the amount that causes toxicity.

Vaccine Hesitancy

Delay in acceptance or refusal of vaccines despite the availability of vaccination services.

Voluntary Licensing

Agreements that allow generic manufacturers to produce patented drugs, improving access in low- and middle-income countries.

WHO Prequalification

A quality assurance process by the World Health Organization that facilitates global procurement of essential medicines.

Annotated Bibliography & Chapter-Wise Notes

PART I: The Foundations of Healing (Pre-World War Era)

Chapter 1: The Birth of Modern Medicine

Notes (Chapter Rationale)

This chapter traces the transition of medicine from empirical tradition to evidence-based pharmacological science. It draws on histories of chemistry, early industrial drug development, and the institutionalization of clinical trials.

Annotated Sources

1. Sneader, Walter. Drug Discovery: A History. Chichester: John Wiley & Sons, 2005.— A definitive scholarly account of how modern drug discovery emerged from chemistry and industrial research. Provides essential context for aspirin, sulfa drugs, and early pharmacology.
2. Porter, Roy. The Greatest Benefit to Mankind: A Medical History of Humanity. London: HarperCollins, 1997.— A classic medical history that situates pharmaceuticals within broader social and scientific change.
3. Weatherall, David. Science and the Quiet Art. Oxford: Oxford University Press, 1995.— Explores the evolution of scientific medicine and the human discipline behind medical breakthroughs.

Chapter 2: Aspirin: The Dawn of Pain Relief

Notes

This chapter combines scientific history with moral reflection, emphasizing aspirin as the first globally standardized drug and a symbol of reproducible medicine.

Annotated Sources

4. Sneader, Walter. "The Discovery of Aspirin: A Reappraisal." BMJ 321, no. 7276 (2000): 1591–1594. Reassesses credit for aspirin's discovery, including Arthur Eichengrün's role.

5. Jeffreys, Diarmuid. Aspirin: The Story of a Wonder Drug. London: Bloomsbury, 2004. Narrative history linking chemistry, corporate development, and public health impact.
6. Bayer Archives. Aspirin—The Brand That Changed Medicine. Leverkusen: Bayer AG, various editions. Used cautiously for historical verification, not corporate endorsement.

Chapter 3:Insulin: A Miracle for the Sugar Disease

Notes

Focuses on insulin as the first successful hormone replacement therapy and a turning point in chronic disease management.

Annotated Sources

7. Bliss, Michael. The Discovery of Insulin. Chicago: University of Chicago Press, 1982. The authoritative historical account of Banting, Best, Macleod, and Collip.
8. Rosenfeld, Louis. "Insulin: Discovery and Controversy." Clinical Chemistry 48, no. 12 (2002): 2270–2288. Explores scientific rivalry and the ethical dimensions of credit for discovery.
9. Diabetes Control and Complications Trial Research Group. "The Effect of Intensive Treatment of Diabetes." New England Journal of Medicine 329 (1993): 977–986. Demonstrates insulin's long-term impact on outcomes.

Chapter 4: Sulfa Drugs: The First Antibiotics

Notes

Examines the transition from fatal infections to treatable conditions and the regulatory awakening following early tragedies.

Annotated Sources

10. Domagk, Gerhard. "A Contribution to Chemotherapy of Bacterial Infections." Deutsche Medizinische Wochenschrift 61 (1935): 250–253. Original scientific report on Prontosil.
11. Lesch, John E. The First Miracle Drugs: How the Sulfa Drugs Transformed Medicine. Oxford: Oxford University Press, 2007. Authoritative historical analysis of sulfa drugs 'social and medical impact.

12. U.S. Food and Drug Administration. The 1938 Food, Drug, and Cosmetic Act. FDA Historical Archive. Establishes the regulatory legacy arising from the sulfanilamide tragedy.

PART II: The Post-War Pharmaceutical Revolution (1940s–1960s)

Chapter 5: Penicillin: The Drug that Won the War

Notes

Positions penicillin as both a scientific breakthrough and a geopolitical turning point.

Annotated Sources

13. Fleming, Alexander. "On the Antibacterial Action of Cultures of a Penicillium." British Journal of Experimental Pathology 10 (1929): 226–236. Foundational discovery paper.
14. Lax, Eric. The Mold in Dr. Florey's Coat. New York: Henry Holt, 2004.Narrative history of penicillin's development and wartime scaling.
15. Bud, Robert. Penicillin: Triumph and Tragedy. Oxford: Oxford University Press, 2007. Balanced account of antibiotic success and resistance.

Chapter 6: Vaccines, Cortisone, and the Age of Wonder Drugs

Notes

Explores the post-war confidence in pharmaceutical science and mass public trust.

Annotated Sources

16. Oshinsky, David. Polio: An American Story. Oxford: Oxford University Press, 2005. Definitive account of vaccine trials and public response.
17. Hench, Philip S. et al. "The Effect of a Hormone of the Adrenal Cortex." Proceedings of the Mayo Clinic 24 (1949): 181–197. Cortisone discovery and clinical impact.

PART III: Pharma in Crisis and Courage

Chapter 7: Thalidomide and the Birth of Drug Safety

Notes

Demonstrates how failure reshaped global regulation.

Annotated Sources

18. McBride, William. "Thalidomide and Congenital Abnormalities." The Lancet 278 (1961): 1358. First public warning on thalidomide.
19. Kelsey, Frances. Protecting the Public Health. FDA Oral History Series. Regulatory courage and reform.

PART IV: The Biotech and Personalized Medicine Era

Chapter 8: From Molecules to Genes

Notes

Tracks the shift from chemical synthesis to biologics and genomics.

Annotated Sources

20. Watson, James. The Double Helix. New York: Atheneum, 1968. Scientific revolution underlying modern biotech.
21. National Institutes of Health. The Human Genome Project. NIH Archives. Foundational reference for personalized medicine.

PART V: Pharma Today and Tomorrow

Chapter 9:

mRNA, Gene Therapy, and the Future Pharma is Building

Notes

Focuses on translational science and platform technologies.

Annotated Sources

22. Dolgin, Elie. "The Tangled History of mRNA Vaccines." Nature 597 (2021): 318–324. Balanced historical account of mRNA development.
23. FDA. Gene Therapy: Scientific and Regulatory Considerations. FDA Briefing Documents.

Further Reading and References

Books

1. The Emperor of All Maladies – Siddhartha Mukherjee, A sweeping history of cancer and the scientists, patients, and doctors who've fought it.
2. The Great Influenza – John M. Barry, A gripping narrative of the 1918 flu pandemic and the emergence of modern medicine.
3. Bottle of Lies – Katherine Eban, An investigative account of the global generic drug industry and the stakes of quality.
4. Gene Machine – Venki Ramakrishnan, A Nobel laureate's insider story of decoding the ribosome and what it reveals about how modern drugs work.
5. The Vaccine Race – Meredith Wadman, Chronicles the birth of modern vaccines and the ethical challenges behind their development.
6. The Innovation Blind Spot – Ross Baird, Explores how systems often overlook promising healthcare innovations and how we can fix it.
7. A Shot to Save the World – Gregory Zuckerman, The inside story of how scientists raced to develop COVID-19 vaccines using mRNA.

Articles & Essays

1. "Why Drug Development Is So Expensive" – Harvard Business Review
2. "Gene Therapy: The Next Frontier" – Nature Reviews Drug Discovery
3. "The mRNA Vaccine Revolution" – Scientific American
4. "The Next 20 Years of Biotech" – MIT Technology Review
5. "How Pharma Partnerships Changed Global Health" – The Lancet Global Health

Reports & White Papers

1. Global Health Estimates – World Health Organization (WHO)

2. State of Vaccine Confidence – Wellcome Trust & LSHTM
3. 2024 Access to Medicine Index – Access to Medicine Foundation
4. Global Tuberculosis Report – WHO
5. Medicines Patent Pool Annual Report – MPP
6. Drug Discovery & Development Roadmaps – IFPMA / BIO / EFPIA

Organizations & Online Resources

1. WHO Prequalification Programme
2. Gavi, the Vaccine Alliance
3. Medicines for Malaria Venture (MMV)
4. Drugs for Neglected Diseases initiative (DNDi)
5. FDA Drug Approval Databases
6. ClinicalTrials.gov – Track ongoing and completed trials

About the Author

Subba Rao Chaganti is a seasoned pharmaceutical professional, thought leader, and prolific author whose work focuses on the intersection of science, marketing, strategy, and ethics in the healthcare industry. With decades of experience in the pharmaceutical business, brand strategy, and industry education, he has dedicated his career to bridging the gap between medical innovation and meaningful human impact.

Chaganti is widely recognized for his efforts to document and explain the evolving role of pharmaceutical science in modern society. His writing spans a broad range of subjects, including pharmaceutical marketing, healthcare strategy, medical innovation, ethics in pharma, and the human stories behind breakthrough medicines. Through his books, articles, and lectures, he aims to make complex industry topics accessible to professionals, students, and informed general readers alike.

His work is distinguished by a balanced perspective that acknowledges both the achievements and the challenges of the pharmaceutical sector while consistently emphasizing the importance of scientific progress, responsible leadership, and patient-centered thinking.

In The Good Pharma, Chaganti brings together his industry insight, historical research, and narrative storytelling to highlight the life-saving contributions of pharmaceutical innovation that have shaped modern civilization.

He continues to write, speak, and mentor on topics related to pharmaceutical strategy, ethical innovation, and the future of healthcare.

Here is a list of his publications:

1. Pharmaceutical Marketing in India: Concepts, Cases, Strategy
2. Game Plans for Post-GATT Era: Action Agenda of Indian Pharmaceutical Industry

3. Compete or Forfeit: Strategies for Sustainable Competitive Advantage in the Pharma Product Patents Era
4. Pharmaceutical Marketing in India for Today and Tomorrow - 25th Anniversary Edition
5. Bullseyes and Blunders: Lessons from 100 Cases in Pharma Marketing
6. Digital Pharma Marketing Playbook: Winning With the New Rules of Engagement
7. Cracking the Generics Code: Your Single-Source Success Manual for Winning in Multi-Source Product Markets
8. Reimagine Pharma Marketing: Make It Future-Proof!
9. Brand Positioning in Pharma
10. Transactional to Transformational Marketing in Pharma: The Science of Why and The Art of How
11. A to Z of Pharmaceutical Marketing: World's First and Only Encyclopedia in Two Volumes
12. The Synergy of Minds: Human+AI Orchestrating the Pharma Marketing Revolution
13. Design Thinking for Pharma
14. The Pharma Product Manager: Navigating the Digital Frontier
15. Sentiment Analysis for Pharma Marketers
16. Whispering to AI: Prompt Engineering for Pharma Brand Managers
17. Storytelling in Pharma Marketing
18. Beyond the Molecule: The Art and Science of Metaphors That Matter
19. The Voices Within: Your Body Speaks When Your Organs Tell Their Own Stories
20. Pharma Archetypes: The Science of Meaningful Healthcare Marketing
21. Navigating the AI Revolution: Human Futures
22. The Wearable Edge: Redefining Pharma Marketing in a Connected and Data-Driven World
23. Code to Cure: Digital Therapeutics and the Future of Pharma
24. The Wearable Edge: Redefining Pharma Marketing in a Connected and Data-Driven World

25. Tarun Gupta: The Man who taught Marketing to the Indian Pharmaceutical Industry
26. The Inner Game of Marketing: Mastering the Mind Behind the Message
27. Level Up: Gamification in Pharma Marketing
28. Marketing Fundamentals
29. The Digital Marketer's Playbook: Strategies, Frameworks, and Tools for the New Marketing Era
30. Advertising in the Age of Generative AI: How Agencies, Creatives, and Brands Win
31. Sardar Patel: The Iron Man of India
32. What Marketers can Learn from Martial Arts: The Path to Mastery and Market Dominance
33. Key Account Management: Strategies for Success

www.ingramcontent.com/pod-product-compliance
Lightning Source LLC
Chambersburg PA
CBHW071223230926
51849CB00001B/12
9788169620567